Optical Networks

Series Editors
Biswanath Mukherjee, Davis, CA, USA
Ioannis Tomkos, Peania, Greece

The book series in Optical Networks and Optical Communication Systems encompasses both optical communications and networks, including both theoretical and applied topics. The series describes current advances at the cutting edge of the field and is aimed especially at industry practitioners, researchers, and doctoral students. The series emphasizes the following major areas:

Area A: Physical Layer Technologies and Subsystems for Optical Transmission and Switching. Topics therein include:

* Optical fibers; Optical amplifier technology; Optical transponder modules; Optical transponder components; Fiber nonlinearities; DSP for optical transponders; FEC for optical transponders; Modulation formats; Advanced techniques for capacity scaling; Wavelength division multiplexing; Space division multiplexing; Optical node architectures; Optical switching approaches; Components for optical switching

Area B: Core Backbone Networks. Topics therein include:

* Optical core network architecture; Routing and wavelength (spectrum) assignment; Flex-grid and elastic optical networks; Resilient/survivable optical networks; Traffic grooming; Dynamic control of optical networks; Cross layer design; Optical network virtualization; Metro networks architectures; Energy efficiency in optical networks; Packet-optical network convergence; Optical packet and burst-switched networks; Cognitive techniques (e.g. AI) in optical network design; Optical network control and management; Software-defined networking

Area C: Access Networks and other applications supporting Human Interaction. Topics therein include:

* Current TDM PON flavors and emerging/advanced PON flavors; Long-reach PON and access/metro integration; Digital optical front-haul technologies and architectures; Analog optical front-haul technologies and architectures; Optical solutions supporting advanced 5G functionalities (e.g. beamforming); Fiber-wireless convergence; Free-space optical links for terrestrial and satellite communication networks; Visible Light Communications for in-house/building networks; Optical communications for avionics and autonomous vehicles; Emerging industry standards

Area D: Optical solutions for Datacenter Networks and High Performance Computing. Topics therein include:

Evolving requirements and trends of datacenters networks; Evolving requirements and trends of HPC; Inter-Datacenter interconnection networks; Intra-Datacenter network architectures; Photonic solutions for DCI and HPC interconnects; System aspects for optical interconnect transceivers; Optical switching for data center networks; Optics in routers/switches; On-board and on-chip optics

The series features the following types of books:

* Intermediate to advanced level textbooks or reference books for graduate and undergraduate courses;
* Expository authored or edited books that extend and unify our knowledge and understanding of particular areas;
* Handbooks and professional reference works that redefine "state-of-the-art," emphasizing expository surveys, completely new advances, and combinations thereof;
* Encyclopedias containing articles dealing with the entire range of knowledge in the field or its specialty subtopics; and
* Authored and edited research monographs that make substantial contributions to new knowledge in specialty subtopics.

For further information, please contact: Biswanath Mukherjee, Series Editor: Distinguished Professor, University of California, Davis, bmukherjee@ucdavis.edu; Ioannis Tomkos, Series Editor: Research Director, AIT, itom@ait.gr; or Mary James, Senior Editor, Springer, mary.james@springer.com.

More information about this series at http://www.springer.com/series/6976

Darli Augusto de Arruda Mello
Fabio Aparecido Barbosa

Digital Coherent Optical Systems

Architecture and Algorithms

 Springer

Darli Augusto de Arruda Mello
School of Electrical & Computer
Engineering
University of Campinas
Campinas, Brazil

Fabio Aparecido Barbosa
School of Electrical & Computer
Engineering
University of Campinas
Campinas, Brazil

ISSN 1935-3839   ISSN 1935-3847   (electronic)
Optical Networks
ISBN 978-3-030-66543-2   ISBN 978-3-030-66541-8   (eBook)
https://doi.org/10.1007/978-3-030-66541-8

This Springer imprint is published by the registered company Springer Nature Switzerland AG
The registered company address is: Gewerbestrasse 11, 6330 Cham, Switzerland

To my parents, Darly and Renée.
Darli

To my parents, Aparecido and Ivone.
Fabio

Preface

This book is the result of a decade of teaching optical communications to graduate and advanced undergraduate students. For many years, we taught optical communications based on the classic books in our area, which are focused on optical systems with intensity modulation and direct detection (IM/DD). We discussed concepts such as chromatic dispersion (CD) maps and reach limitations caused by CD and polarization mode dispersion (PMD). However, digital coherent techniques profoundly changed the optical communications landscape. We soon realized that this basic course in optical communications was becoming uninteresting to many students already working in the field. Discussions such as CD management, and reach limitations due to CD and PMD, no longer were applicable to practical systems. On the other hand, merging in a one-semester course, the introductory content of optical communications, and fundamentals of coherent systems, seemed challenging. This was exactly the inspiration for this book.

The purpose of this book is to support a one-semester course in optical communications, providing introductory content and still covering the main digital signal processing (DSP) algorithms in coherent optical systems. The course can be taught to graduate students, or advanced undergraduate students with background on digital communications and electromagnetic theory. Compared with classical optical communications courses, we remove much of the content relevant to systems with IM/DD, but less important for modern coherent systems. In order to accelerate the study of DSP algorithms, we present off-the-shelf Matlab/Octave functions within the book text. It remains for the student, therefore, to understand the presented concepts and build his or her own simulator based on the available functions. The book can also serve as an auxiliary reference to a basic or advanced optical communications course, covering the most relevant DSP algorithms.

We express our appreciation to the students who helped us to test the problems, Luan dos Santos, Robson Colares, Rafael Alvarenga, Seyed Mahjour, Rafaela Cardoso, and Felipe Sanches. We owe a debt of gratitude to Cristiano Panazio, Hélio Waldman, Joseph Kahn, Lucas Gabrielli, Nebojsa Stojanovic, Renato Rocha Lopes, and Tobias Fehenberger, who helped us to review parts of the manuscript.

Our special thanks to Valery Rozental, Thiago Portela, Lailson dos Santos, Hugo Ferreira, and José Hélio Junior, whose theses helped to structure this book. Finally, our special thanks to Ioannis Tomkos, who greatly encouraged this project.

Campinas, Brazil   Darli A. A. Mello
Campinas, Brazil   Fabio A. Barbosa
October 2020

Contents

Chapter 1
Introduction

1.1 The Invention of Digital Coherent Optical Systems

This book is entitled *digital coherent optical systems*. But, in fact, there is a bit of imprecision in this nomenclature. Perhaps it would be most correct to name it *digital optical communication systems with coherent detection and digital signal processing*, but this would be a little long. But it is now common to refer to such systems as simply "digital coherent optical systems." Coherent detection in optical systems dates back to the 1980s when it was used to increase spectral efficiency and sensitivity. Actually, most of the technologies that form the basis of the current coherent optical systems were already present at that time [1–3]. However, with the astounding multiplication of capacity promoted by wavelength division multiplexing (WDM) in the 1990s, it stayed in the background for a long time, more precisely until the mid-2000s. And it came back due to a conjunction of technologies that have become the heart of modern optical communications.

In order to understand the reasons that led to the revival of coherent detection, let us take a time-back exercise in the early 2000s. The advent of the Internet had caused a great excitement in the technology markets, and promising ideas had the potential to change the world. It would be the beginning of something like a new industrial revolution that would change forever the way we would relate. Looking back at 20 years ago, one realizes that this worldview was correct, but that this evolution (or revolution) was not at all linear, and went through stumbles that shook the world of technology. Returning to our time travel exercise, in the start of 2000s our community was full of intense excitement. With prospects of exponential increase in data traffic, large investments had been made to increase the data rate of optical communication systems. But this excitement of the early 2000s soon gave way to the frustration of the Internet bubble burst. Although the demand for traffic kept increasing in a good pace, the growth rate was less than expected. Technology companies shut down all over the world, and the optical communications industry suffered like few others.

© Springer Nature Switzerland AG 2021
D. A. de Arruda Mello, F. A. Barbosa, *Digital Coherent Optical Systems*,
Optical Networks, https://doi.org/10.1007/978-3-030-66541-8_1

At that time, the 10 Gb/s transponders were beginning to be commercialized, and the new studies focused on future 40 Gb/s systems. One thing seemed right, it would be very difficult, if not impossible, to transmit 40 Gb/s channels over the same infrastructure as the 10 Gb/s systems. At these rates, moderate values of polarization mode dispersion (PMD) and residual chromatic dispersion (CD) are highly deleterious to transmission. The CD parameter D has unit of ps/nm/km, and a measure of the time spread of a pulse in ps is given by D multiplied by the spectral width in nm and the fiber length in km. Thus, if the bit rate increases by a factor of 4, the pulse spread also increases by a factor of 4, following the increase in signal bandwidth. As the pulse duration also decreases by a factor of 4, the overall tolerance to the CD parameter decreases by a factor of 16, i.e., with the square of the symbol rate. Fortunately, the accumulated CD in the end of the link can be fine-tuned by optical dispersion compensation techniques. The situation of links with PMD was more dramatic. The PMD tolerance would decrease in inverse proportion to the increase in bit rate. However, as some links were already severely affected by PMD, further decreasing the tolerance by a factor of 4 would be unacceptable. While optical compensators for PMD had been developed and were effective, they were considered too expensive to deploy. Each receiver would need to be equipped with its own adaptive optical PMD compensator, owing to the fact that PMD changes on the millisecond or sub-millisecond time scale and varies significantly from one WDM channel to another. In addition, binary signals with intensity modulation have a spectrum width that can reach twice the bit rate, hampering propagation in the current 50-GHz grid. Thus, much of the research focused on optimizing the systems to levels never before seen. Spectrally efficient modulation formats for that time such as duobinary (DB) or differential quadrature phase-shift keying (DQPSK) quickly gained the interest of the community. Spectral efficiency (SE) has several definitions, and one is the ratio of information rate in bit/s to the spectral width used to convey the information in Hz. Thus, SE can be measured in b/s/Hz. With on–off keying (OOK), the spectral efficiency of 10 Gb/s transmitted over the 50-GHz grid is 0.2 bit/s/Hz, and with 40 Gb/s, it increases to 0.8 bit/s/Hz. A few years later systems with coherent detection at 100 Gb/s were operating with spectral efficiencies reaching 2 bit/s/Hz using the legacy 10 Gb/s infrastructure.

Behind this surprising twist is the ingenuity of engineers who have brought coherent detection back into the game, but in a completely different outfit, full of innovations with an irresistible performance. Until that moment, most receivers used direct detection based on a single photodetector. The photodetector converts the electric field of the optical signal $E(t)$ into electric current $i(t)$ by a simple relation $i(t) = R|E(t)|^2$, where R is a proportionality factor known as photodetector responsivity. Any phase information in $E(t)$ is lost in the photodetection process. This nonlinear mapping between the received optical field and the detected current limited those optical systems in two ways. The first is to rule out phase modulation of the optical carrier, which clearly limits the spectral efficiency of the system. Second, the nonlinearity of the process precludes the use of robust linear filtering techniques for CD and PMD compensation. Initiatives to increase spectral efficiency began to be carried out through the use of differentially encoded phase modulation

[4], using delay interferometers at the receiver to convert phase changes into intensity changes.

The first signs of a return to coherent detection appeared in 2004. Taylor demonstrated in [5] an experimental setup that would be widely used thereafter in laboratory experiments, consisting of a front-end responsible for converting the received optical signal into electric signals, followed by a high-speed oscilloscope capable of storing data windows for offline processing. The paper also demonstrated a digital phase recovery algorithm without the need of a phase-lock loop, in addition to compensating for a relatively high amount of CD in the digital domain. At the same time, a synchronous receiver with in-phase and quadrature baseband processing was proposed by Noé in [6]. These papers indicated a return to the coherent detection concepts of the 1990s, but now with greater interdependence with digital signal processing (DSP) algorithms. However, there were still a number of technical and commercial obstacles to overcome. These early works seemed to be able to resolve the issue of CD and phase synchronization but still lacked a more robust control of the effects of polarization, such as PMD.

It was not long before this issue was also addressed using DSP concepts. Noé proposed in [7] a DSP-based electronic polarization control. Han and Li suggested in [8] that optical communication systems using polarization multiplexing would be analogous to wireless communications systems using MIMO, and that algorithms for channel estimation in wireless MIMO could be applied to separate signals multiplexed in polarization. In the following years, digital coherent optical systems gained maturity through contributions from various actors [9–13], but there were still many doubts about its commercial viability. These doubts quickly faded with the implementation of the first 40 Gb/s transponders with coherent detection and real-time processing demonstrated by Nortel in 2008 [14]. Building on the work published in the previous years and relying on a multidisciplinary team of more than 100 people from diverse areas, Nortel's 40 Gb/s transceiver definitively consolidated coherent detection as the new standard for optical systems at high data rates. One of the key factors for the success of the team was the development of a miniaturized analog-to-digital converter (ADC) capable of operating at the rates required for these systems. Suddenly, 40 Gb/s systems, which previously had low CD tolerance, could now completely remove optical dispersion compensators from the link. PMD, the great villain of optical systems with intensity modulation and direct detection, was now fully compensated by an adaptive digital filter. The complete removal of dispersion compensation modules improved the quality of the link because of reduced insertion losses and increased tolerance to nonlinear effects. The digital signal processors at the transmitter and receiver proved to be excellent platforms for innovation, where research groups around the world presented their latest inventions. Certainly, a new generation of optical communication systems had been inaugurated.

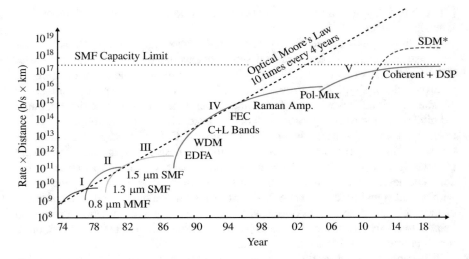

Fig. 1.1 Rate-distance (BL) product per fiber evolution [15, 16]: Roman numerals indicate the system generation; MMF—multimode fiber; SMF—single-mode fiber; EDFA—erbium-doped fiber amplifier; WDM—wavelength division multiplexing; FEC—forward error correction; Raman Amp.—Raman amplification; Pol-Mux—polarization multiplexing; DSP—digital signal processing; SDM*—space-division multiplexing (as of the writing of this book, SDM systems over multicore or multimode fiber have not yet become commercially available)

1.2   How Did We Get Here?

So far we have discussed how the ingenuity of scientists working in academia and industry fostered a revolution in optical communications through the introduction of digital coherent optical systems. However, this was not the beginning of the story, nor the end of it. In fact, digital coherent optical systems are the fifth generation of optical systems, and the development of the sixth generation is already underway. Thus, we present here a brief history of optical communications and the technological perspectives that are yet to come.

Most introductory texts on optical communications start with the discussion shown in Fig. 1.1, which depicts the evolution of the rate-distance product, or BL product, over the time [15, 16]. Optical communication systems are known to have a large bandwidth, and this bandwidth can be used to transmit large volumes of information over large distances. Thus, the BL product is the most suitable metric to quantify the quality of a fiber optic link and the capabilities of different technological generations. One interesting feature in Fig. 1.1 is the shape of the curves corresponding to different generations. Each curve begins with a large derivative that then decreases over time. This great derivative corresponds to the introduction of disruptive technologies, with the potential to revolutionize the state of the art in a short time. These disruptive technologies are then followed by incremental contributions that are equally important to the evolution of technology.

The first considerations in using glass fibers for long-distance transmission have been made by C. K. Kao at Standard Telecommunications Laboratories Ltd. (STL), a subsidiary of ITT Corporation in the UK. Together with G. A. Hockham, he published in 1966 the paper entitled "Dielectric-fibre surface waveguides for optical frequencies," which is today recognized as the birth of fiber-optic communications [17]. Their invention would come to transform the world and give Kao the 2009 Nobel Prize in Physics. The paper already contained the main elements of today's optical fibers, including cylindrical geometry, with a core of higher refractive index surrounded by a material of lower refractive index. But there was one essential problem: the relatively low level of glass purity and its high losses. With the technology of the time, the glass attenuation coefficient was around 200 dB/km [18]. Even so, in their visionary work, Kao and Hockham suggested that attenuation coefficients in the order of 20 dB/km could be achieved. The invention of Kao and Hockham was then improved in the following years, mainly aiming at the purification of glass and the reduction of its losses. Significant advances were made by the research laboratories at Corning Glass, where Kapron, Keck, and Maurer announced losses below 20 dB/km in single-mode fibers hundreds of meters long [19, 20]. As we know today, the 20 dB/km estimate of Kao and Hockham was quite conservative, and the current standard single-mode fibers (SMFs) achieve attenuation coefficients below 0.2 dB/km. These levels of attenuation were reached as early as the end of the seventies by the NTT researchers Miya, Terunuma, Hosaka, and Miyamashita, in [21].

This first generation of optical communication systems used AlGaAs light-emitting diodes (LEDs) and lasers at 0.8 μm operating in graded-index multimode fibers. Limited by modal dispersion, these systems reached several hundreds of Mb/s·km [22, 23]. The development of SMFs in the early eighties overcame the problem of modal dispersion, inaugurating a second generation of optical fiber communication systems. In addition, the development of new InGaAsP sources moved the operating regime toward the so-called long wavelengths, in the window near 1.3 μm, where the fiber attenuation is lower [24, 25]. The third generation of the mid-eighties moved the operating wavelength window from 1.3 μm to 1.5 μm, further reducing the attenuation of the optical fiber channel [26, 27]. The fourth generation of the late eighties promoted a quantum leap in optical communications and enabled a new era of low-cost data transmission. With the development of single-mode systems operating at 1.5 μm, the main limitation of optical transmission systems moved from the optical fiber to the electronics [27], a situation known as electronic bottleneck. Specifically, although the fiber had enough unexploited bandwidth, it remained unused because the electronics could not modulate and detect signals at those rates. The natural solution to this problem is to use frequency division multiplexing, in which different signals share the same fiber, using different portions of the spectrum.

The trend in using frequency division multiplexing or, in optical communications, WDM is explained in Fig. 1.2a and b. Figure 1.2a shows the configuration of third-generation systems, in which transmitter and receiver were interconnected by a single optical signal transmitted over a single fiber. In this configuration, increasing

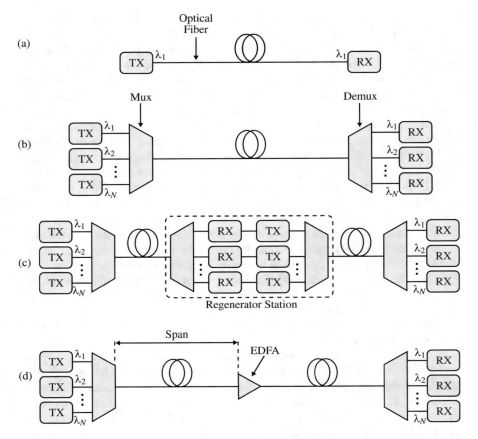

Fig. 1.2 (**a**) Single-wavelength unamplified optical transmission system (first to third genera-tions), consisting of transmitter, fiber, and receiver. (**b**) WDM unamplified short-reach system. Several optical signals operating at different wavelengths are wavelength-multiplexed using a multiplexer, and wavelength-demultiplexed using a demultiplexer. The system data rate is multiplied by the number of operating wavelengths. (**c**) WDM unamplified long-reach system. The optical signals must be terminated and regenerated in the electrical domain after the maximum span distance (typically, in the order of 100 km). The regeneration cost scales linearly with the number of supported wavelengths. (**d**) WDM long-reach optical system with EDFA amplification (fourth generation). Compared with (**c**), the cost of the solution is strongly reduced, as regenerator stations are replaced by a single EDFA

the link data rate would require the activation of additional fibers. Using WDM, the outputs of several transceivers operating in different wavelengths are optically multiplexed into a single fiber using a multiplexer (see Fig. 1.2b). At reception, the wavelength-multiplexed signals are separated using a demultiplexer. In principle, this solution solves the electronic bottleneck, but it is still not scalable in terms of cost. As the fiber attenuates the signal, it needs to be regenerated periodically in the electronic domain to be able to recover the transmitted bits. The problem of this approach is that opto-electro-optical converters are expensive, and each operating wavelength requires its own regenerator (see Fig. 1.2c). As a consequence, the cost of a regenerator station becomes proportional to the number of active wavelengths. As the distance between two regeneration stations typically does not exceed 100 km, the cost of long-haul optical links with thousands of kilometers and tens of regeneration stations was prohibitive.

The solution to this problem came with the invention of an ingenious and relatively inexpensive device, capable of amplifying all wavelengths in the optical domain, without any opto-electro-optical conversion. The erbium-doped fiber amplifier (EDFA) revolutionized not only the optical communications, but also made possible the information society in which we live today. Nowadays, we pay the same amount to transmit a bit to a neighboring city, as to a server on the other side of the globe. This is not the case with telephone calls or with freight. In fact, usually the price of transporting a commodity is proportional to the distance to be traveled. The inversion of this logic in information systems was only possible by the invention of the EDFA, for which the incremental cost of data rate is extremely low. The invention of the EDFA is usually attributed to two groups at the University of Southampton in the UK, and at Bell Labs in the USA. The group at Southampton led by Prof. D. N. Payne pioneered the investigation of optical properties of rare-earth-doped fibers [28] and, after a few years, demonstrated the first EDFAs. In [29], Mears et al. gave the first signs of what was to come: *"There are currently two competing technologies for optical repeaters, namely semiconductor laser amplifiers and Raman amplifiers. In this letter we demonstrate a third approach, based on an optically pumped rare-earth-doped optical fibre."* These early demonstrations were quickly improved by the efforts led by E. Desurvire at the AT&T Bell Laboratories [30]. The importance of the invention of the EDFA can be understood by comparing Fig. 1.2c and d. Prior to the invention of EDFA, even with WDM, the optical signal for each channel (or wavelength) needed to be periodically regenerated in the electronic domain to compensate for link losses. This regeneration needed to be performed at distances of less than 100 km. Therefore, a WDM link of 20 wavelengths and 1000 km required $9 \times 20 = 180$ opto-electro-optical regenerators. With the advent of EDFA, these 180 regenerators could be replaced by only 9 EDFAs, with each EDFA having a lower cost than a single regenerator. The cost savings are even more dramatic in ultra-long-haul and transoceanic distances with hundreds of channels. As with other technologies, incremental innovations were followed, including the implementation of error-correcting codes, C+L band transmission, and the deployment of hybrid EDFA/Raman amplification.

The fourth generation of optical systems lasted until the mid-2000s, when the technological revolution enabled by the Internet, the expansion of mobile equipment, and the popularization of cloud services required the development of a new disruptive technology. It is in this context that the digital coherent systems appeared, inaugurating the fifth generation of optical transmission systems, which we discussed in the previous section. As of today (year 2020), the highest reported

BL product over an SMF is in the order of 800 Pb/s·km [31]. The sixth generation of optical systems is yet to come, and it may be premature to say with certainty the technology to be employed. But the current state of the art points to the development of technologies that exploit space-division multiplexing (SDM), either by implementing fibers with multiple uncoupled cores, or by the transmission of coupled signals, but later separating them by digital signal processing (DSP) with multiple outputs and multiple inputs (MIMO) [32–34]. Recent SDM demonstrations reach BL products in the order of a few Eb/s·km [35–38].

1.3   Digital Coherent Optical Systems in Modern Optical Networks

So far we have discussed the evolution of point-to-point optical systems toward ever-increasing reaches and data rates. And also the remainder of this book delves exclusively into point-to-point transmission technologies. However, it is important not to lose sight of the perspective that, between generation and detection, there are several network elements that route the optical signal from a source node to a destination node. Therefore, this section briefly describes the main network elements in digital coherent optical systems.

1.3.1   From Routers to Wavelengths

Digital coherent optical systems are mainly deployed in high data rate core optical networks intended to provide Internet traffic between geographically distant locations. These sites usually belong to corporations that use the Internet for various purposes, such as Internet access providers, banks, and cloud service providers. The vast majority of signals transmitted in today's optical networks originate from Ethernet-based routers, which aggregate various low-rate streams into a high-rate stream, and launch it into an optical network, as shown in Fig. 1.3a. The generated optical signal operates in the windows of 850 nm or 1310 nm and is transmitted over SMFs or (low-cost) multimode fibers. This short-reach optical signal is sent to a device called *transponder*, whose purpose is to receive the optical signal generated by a low-cost laser (colorless port) and prepare the information for long-distance transmission. This preparation includes the encapsulation of information in a transport protocol (e.g., the optical transport network architecture specified in the ITU G.872 recommendation), the insertion of forward error correction (FEC), and optical domain conversion using a high-quality (and high-cost) laser operating at the 1550 nm window (colored port).

You may be wondering why the router cannot itself generate the signal that will be routed through the optical network. In principle it can, and this is an issue

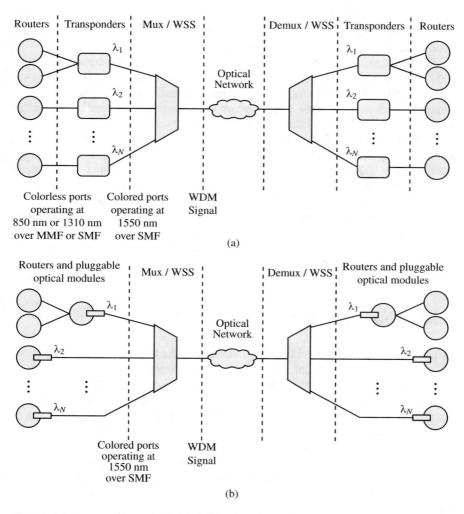

Fig. 1.3 (**a**) Transponders and WDM signal generation. The routers are the main clients of optical networks. They generate short-reach optical signals operating at the windows of 850 or 1310 nm (colorless ports). Transponders receive these signals and include additional overheads corresponding to control and forward error correction (FEC). Transponders also generate high-quality optical signals at the window of 1550 nm (colored ports). These signals are subsequently wavelength-multiplexed by a fixed-grid multiplexer or WSS. Finally, the signal is launched into the optical network. The receiving node carries out analogous operations in reverse order. (**b**) In the IP/WDM configuration, transponders are replaced by pluggable optical modules with colored ports

intensively debated in our community. For operational and commercial reasons, it has been often preferred to decouple the optical equipment from the routers. Among the advantages of this separation is the possibility of using different suppliers for the two layers, which tends to increase competitiveness and reduce the cost of the solution. Also, sometimes the transponder multiplexes sub-rate data from several routers, implementing a *muxponder* functionality.

The use of transponders is an active issue of discussion, and the common practice may change with the evolution of technology. The Internet Protocol (IP) over WDM (IP/WDM) architecture, where routers are directly connected to WSSs using colored interfaces (see Fig. 1.3b), is becoming increasingly popular in the context of software defined networking (SDN). In the SDN paradigm, all network elements have standardized interfaces that allow them to be controlled by a single centralized control plane. Data and optical equipment from different vendors become "white boxes" whose main functionalities are managed by the control plane. This strategy is particularly attractive for data center interconnection networks, which are the primary bandwidth consumers at the present time. A second trend that is gaining momentum is the co-packaging of networking and DSP functions in a single chip. As the rates of optical systems increase substantially, and optical transceivers become increasingly DSP intensive, electronic co-packaging may avoid intra-device opto-electro-optical conversion.

1.3.2 Wavelength Multiplexing and the Flexible Grid

After the high-quality optical signal is generated by the transponder, it is frequency-multiplexed using an optical multiplexer. At the receiver, the reverse process is carried out by an optical demultiplexer. Multiplexers and demultiplexers are devices of the same type but operated with reverse input and output ports. They can be passive devices constructed using arrayed waveguide gratings (AWGs) [15, 39], or the complex wavelength selective switches (WSSs). We will discuss the operation of WSSs later on in this section. The transponder is tuned to generate an optical signal at a certain wavelength that remains fixed up to the receiver. We call this logical entity between transmitter and receiver a *lightpath* [40]. Until the mid 2000s, each lightpath occupied a 50-GHz bandwidth independently of its rate (2.5 or 10 Gb/s). The signals generated at transponders were tied to multiplexer and demultiplexer ports corresponding to specific frequencies of a channel grid.

It is worth mentioning here that, if channels are equally spaced in frequency, they are not equally spaced in wavelength. For example, a certain channel with frequency f_1 corresponds to a wavelength $\lambda_1 = c/f_1$, where c is the light velocity in vacuum. Another channel with frequency f_2 corresponds to a wavelength $\lambda_2 = c/f_2$. A channel spacing in frequency is given by $f_2 - f_1 = c/\lambda_2 - c/\lambda_1 = c\left(\frac{\lambda_1 - \lambda_2}{\lambda_2 \lambda_1}\right)$. Assuming that channels 1 and 2 are sufficiently close in spectrum, $\lambda = \lambda_1 \approx \lambda_2$, and $f_2 - f_1 \approx \frac{c}{\lambda^2}(\lambda_1 - \lambda_2)$. Thus, at $\lambda = 1550$ nm, which is the region of lowest attenuation in an optical fiber, a frequency spacing of 50 GHz corresponds to a wavelength spacing of approximately 0.4 nm.

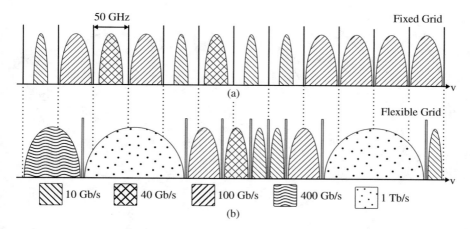

Fig. 1.4 Fixed and flexible grid. (**a**) In fixed-grid systems, all optical channels are allocated over the same bandwidth. If the channel spacing is fixed at 50 GHz, spectrum is wasted for lower-rate channels (e.g., 10 Gb/s), and higher-rate channels (e.g., 400 Gb/s) are not supported. (**b**) A flexible grid can accommodate higher- and lower-rate channels with optimized spectrum allocation

With the evolution of technology, and the development of transponders operating at different rates, the fixed frequency grid became inefficient. While low-rate connections needed a spectrum of less than 50 GHz (2.5 or 10 Gb/s), other higher-rate connections (e.g., reaching rates in the order of 400 Gb/s) needed more than 50 GHz. In parallel, the development of WSSs allowed to allocate optical channels of different bandwidths. This combination of technological advances and commercial needs favored the rapid adoption of a flexible grid [41]. In response to this new scenario, the ITU-T standard G.694.1 of 2012 allowed channel spacings of multiples of 12.5-GHz frequency slots. Figure 1.4 illustrates the differences between a fixed grid and a flexible grid. In Fig. 1.4a, 10 Gb/s and 100 Gb/s channels coexist in a 50-GHz fixed-grid network. Channels at 400-Gb/s are not supported. In Fig. 1.4b, channels with rates from 10 Gb/s up to 1 Tb/s coexist with optimum spectrum allocation. With the fixed grid, the 10 Gb/s channels occupy a 50-GHz band. With the flexible grid, two 12.5 GHz slots would be sufficient, saving 50% of the available bandwidth. In practice, 400 Gb/s and 1 Tb/s channels may require a varying number of frequency slots, depending on the modulation format and coding scheme [42–44].

1.3.3   Wavelength Routing

Until the mid-1990s, optical communication systems were exclusively used to transport voice traffic generated by telephone networks. Telephone traffic has a profile that favors connections between neighboring users, being highly suitable for ring topologies, as the one shown in Fig. 1.5a. However, the advent of the

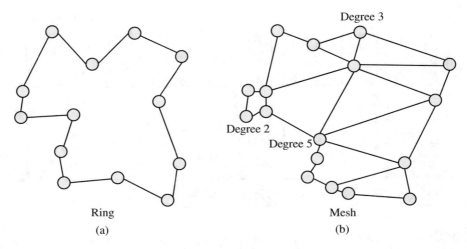

Fig. 1.5 Optical network topologies. (**a**) In the 1990s, voice traffic was dominant and optical systems were operated with point-to-point or ring topologies. (**b**) With the advent of the Internet, mesh networks became the standard. Nodes in mesh networks can be characterized by their degree, which is the number of neighboring nodes. The higher the node degree, the more complex the node switching architecture

Internet in the mid-1990s changed this logic [45]. For an Internet user, it is more likely to access Internet content in a different city, or even in a different country, than in neighboring areas. As a consequence, optical networks evolved from ring and point-to-point topologies in the 1980s to mesh topologies in the 2000s. The evolution in network technologies in the mid-2000s had a symbiotic relationship with the development of digital coherent optical systems, and the combination of transmission and networking techniques built a much more efficient optical network [39]. Therefore, most current optical communication systems have mesh topologies, as the one shown in Fig. 1.5b.

The nodes of a mesh topology can be characterized by their degree, which is the number of neighboring nodes. For example, in a ring topology, all nodes have degree two. A mesh topology is defined as such by the existence of several interconnected nodes with degree greater than two. Lightpath forwarding is carried out by network elements called reconfigurable add-drop multiplexers (ROADMs). The main building block in ROADMs is wavelength selective switches (WSSs), whose logical operation is exemplified in Fig. 1.6 [46]. Working as a multiplexer (see Fig. 1.6a), the WSS is able to select any set of wavelengths from any of its input ports and steer it to the output port. Working as a demultiplexer (see Fig. 1.6b), the WSS is able to select any set of wavelengths from its input port, and steer it to any of its output ports. The physical operation of a WSS working as a multiplexer is shown in Fig. 1.6c. The input port receives a WDM signal and spatially separates the constituent wavelengths using a diffraction grating. The demultiplexed signals are then projected onto a surface that reflects the incident signals and steers them

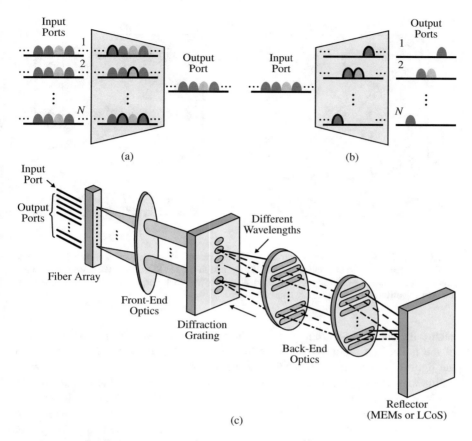

Fig. 1.6 (**a**) Logical operation of a WSS working as a multiplexer. The WSS selects any set of wavelengths from any input port and steers it to its output port. (**b**) Logical operation of a WSS working as a demultiplexer. The WSS selects any set of wavelengths from its input port and steers it to any of its output ports. (**c**) Physical architecture of a WSS working as a demultiplexer. After the front-end optics, an input WDM signal is wavelength-demultiplexed by a diffraction grating. The back-end optics direct the several wavelengths to a mirror able to spatially direct each wavelength to the desired output port

in the direction of the desired output diffraction grating. This reflector can be constructed on microelectromechanical machines (MEMs) [47] or liquid crystal on silicon (LCoS) technologies [48]. The output diffraction grating (one per output fiber) multiplexes the input WDM signals and sends them to the corresponding output port.

The combination of multiple WSSs and eventually power splitters enables the design of the two main ROADM architectures [49] shown in Fig. 1.7. In the *route and select* architecture (R&S), shown in Fig. 1.7a, there is one WSS for each input fiber, and one WSS for each output fiber. In total, the ROADM has 2N WSSs,

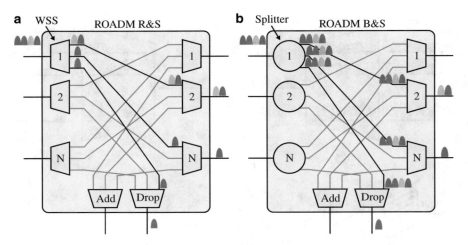

Fig. 1.7 (**a**) R&S ROADM architecture. WSSs are placed at input fibers and at output fibers. R&S architectures avoid excessive power splitting in high-degree ROADMs. (**b**) B&S architecture. Power splitters are placed at input fibers, and WSSs are placed at output fibers. R&S architectures are cost effective and reduce narrowband filtering

each with one input–output port, and N output–input ports, being N the degree of the node. The input WSS directs the desired lightpath (indicated by colors in the figure) from a certain input fiber to the desired output fiber. The output WSS multiplexes the desired lightpaths from several input fibers into the output fiber. In the *broadcast and select* (B&S) architecture, shown in Fig. 1.7b, there is one 1:N power splitter for each input fiber, and one WSS for each output fiber, each with N input ports and one output port, where N is the degree of the node. The power splitter replicates the WDM input signal and feeds it to all output WSSs. The output WSS then selects the desired portion of the spectrum to be transmitted. The R&S architecture usually outperforms the B&S architecture, as it offers lower insertion losses for large node degree N. On the other hand, the B&S architecture avoids extra filtering and polarization-dependent loss. Furthermore, by saving N WSSs, the B&S becomes an interesting solution from an economical standpoint. In addition to the express ports, shown on the left and right sides of Fig. 1.7, there are also add and drop ports, shown at the bottom of the figure. The role of add and drop ports is to insert lightpaths into the source node (add port) and remove them at the destination node (drop port).

Among several desired properties, ROADMs are sometimes required to be *colorless*, *directionless*, and *contentionless* (CDC). A ROADM is called colorless when all of its add and drop ports can operate at any input and output wavelength selected by the network operator. This is achieved by the use of WSSs for adding and dropping channels, instead of typical multiplexers and demultiplexers, for which output and input ports are tied to specific wavelength channels. In a directionless architecture, the input wavelength arriving from any input node can be remotely directed to any output node by the network operator. This property is achieved by both configurations shown in Fig. 1.7. Finally, the contentionless property is perhaps the most challenging one. If the ROADM is contentionless, two (or more) lightpaths having the same wavelength can be added or dropped at the same time and directed to different neighboring nodes. Analogously, a contentionless ROADM is able to drop two (or more) lightpaths at the same time and having the same wavelength but originated from different neighboring nodes. A contentionless architecture requires complex add and drop modules consisting of multiple WSSs, which can be excessively expensive for the degree of flexibility added by this feature. In fact, with suitable routing and wavelength assignment algorithms, the performance of networks with colorless and directionless ROADMs can be made similar to that of networks using CDC ROADMs [50]

Current optical systems offer flexibility in several dimensions, and this is only possible thanks to symbiotic advances in transmission (with coherent detection) and networking (with flexible-grid WSSs and ROADMs) [44]. In this section, we studied how lightpaths are routed by ROADMs over the optical network until reaching the coherent detector. In the following chapters, we will abstract these networking issues and focus exclusively on the generation, transmission, and detection of optical signals. We will also give up the discussion of systems with intensity modulation and direct detection and their devices to focus solely on digital coherent optical systems.

1.4 Problems

1. A 10 Gb/s optical receiver with OOK modulation tolerates $\pm$ 600 ps/nm accumulated CD. Calculate the CD tolerance for an equivalent transceiver operating at 40 Gb/s.
2. An optical link has an accumulated CD of 700 ps/nm. The transceiver operates with OOK 10 Gb/s and has $\pm$ 600 ps/nm accumulated CD tolerance. Calculate the range of accumulated CDs in dispersion compensation modules (DCMs) required to bring the link accumulated CD to values acceptable to the transceiver. Repeat the problem assuming OOK 40-Gb/s transceivers.
3. An optical link has a PMD of 1 ps/$\sqrt{\text{km}}$. Calculate the maximum length allowed by PMD, if an OOK transceiver operates at 10 Gb/s, and its delay spread tolerance is 10% of the symbol time. Repeat the problem for 40 Gb/s.
4. Assume that the bandwidth of a signal with OOK modulation is twice the symbol rate. Calculate the maximum spectral efficiency (in b/s/Hz) achievable by the system. Calculate the spectral efficiency of a 10-Gb/s system in a 50-GHz grid. Would an OOK 40-Gb/s signal fit into a 50-GHz grid?
5. An optical system has a BL product of 10^{10} bit/s·km. Does it correspond to which generation of optical transmission systems? Describe the main properties of its generation. What is the maximum achievable bit rate in a 100-km link? Which is maximum achievable reach with 1 Gb/s?

6. Calculate the approximate channel width in [nm] for a 50-GHz channel in the windows of 1500 nm and in 850 nm.
7. An optical network operates with a fixed 50-GHz grid. The network supports traffic with a uniform mix (with the same number of channels) of 10-Gb/s channels with OOK modulation occupying a 20-GHz bandwidth, and 100-Gb/s channels with multilevel modulation and coherent detection, occupying a 50-GHz bandwidth. Calculate the improvement in spectral efficiency (in b/s/Hz) by using a flexible grid with 12.5-GHz bandwidth resolution, with respect to the fixed grid with 50-GHz bandwidth.
8. Calculate the total number of WSSs and the number of ports in each WSS required to implement the mesh topology depicted in Fig. 1.5. Neglect the WSSs used for add and drop. Assume the R&S ROADM architecture in Fig. 1.7. Repeat the problem with the B&S ROADM architecture in Fig. 1.7. Take into account that the network consists of nodes with different degrees.

References

1. L.G. Kazovsky, Phase- and polarization-diversity coherent optical techniques. J. Lightwave Technol. **7**(2), 279–292 (1989)
2. J.R. Barry, E.A. Lee, Performance of coherent optical receivers. Proc. IEEE **78**(8), 1369–1394 (1990)
3. F. Derr, Coherent optical QPSK intradyne system: concept and digital receiver realization. J. Lightwave Technol. **10**(9), 1290–1296 (1992)
4. K.-P. Ho, *Phase-Modulated Optical Communication Systems* (Springer, Boston, MA, 2005)
5. M.G. Taylor, Coherent detection method using DSP for demodulation of signal and subsequent equalization of propagation impairments. IEEE Photon. Technol. Lett. **16**(2), 674–676 (2004)
6. R. Noé, Phase noise tolerant synchronous QPSK receiver concept with digital I&Q baseband processing, in *Proceedings of Opto-Electronics and Communications Conference (OECC)*, Yokohama, Japan, Jul. 2004, pp. 818–819
7. R. Noé, PLL-free synchronous QPSK polarization multiplex/diversity receiver concept with digital I&Q baseband processing. IEEE Photon. Technol. Lett. **17**(4), 887–889 (2005)
8. Y. Han, G. Li, Coherent optical communication using polarization multiple-input-multiple-output. Optics Express **13**(19), 7527–7534 (2005)
9. E. Ip, J.M. Kahn, Digital equalization of chromatic dispersion and polarization mode dispersion. J. Lightwave Technol. **25**(8), 2033–2043 (2007)
10. E. Ip, J.M. Kahn, Feedforward carrier recovery for coherent optical communications. J. Lightwave Technol. **25**(9), 2675–2692 (2007)
11. D.S. Ly-Gagnon, S. Tsukamoto, K. Katoh, K. Kikuchi, Coherent detection of optical quadrature phase-shift keying signals with carrier phase estimation. J. Lightwave Technol. **24**(1), 12–21 (2006)
12. S.J. Savory, G. Gavioli, R.I. Killey, P. Bayvel, Electronic compensation of chromatic dispersion using a digital coherent receiver. Optics Express **15**(5), 2120–2126 (2007)
13. S.J. Savory, Digital filters for coherent optical receivers. Optics Express **16**(2), 804–817 (2008). [Online]. Available: http://www.opticsexpress.org/abstract.cfm?URI=oe-16-2-804
14. H. Sun, K.-T. Wu, K. Roberts, Real-time measurements of a 40 Gb/s coherent system. Optics Express **16**(2), 873–879 (2008). [Online]. Available: http://www.opticsexpress.org/abstract.cfm?URI=oe-16-2-873
15. G.P. Agrawal, *Fiber-Optic Communication Systems*, 4th edn. (Wiley, Hoboken, NJ, 2010)

16. E. Desurvire, C. Kazmierski, F. Lelarge, X. Marcadet, A. Scavennec, F. Kish, D. Welch, R. Nagarajan, C. Joyner, R. Schneider, Science and technology challenges in XXIst century optical communications. C. R. Phys. **12**, 387–416 (2011)
17. K.C. Kao, G.A. Hockham, Dielectric-fibre surface waveguides for optical frequencies. Proc. Inst. Electr. Eng. **113**(7), 1151–1158 (1966)
18. C.K. Kao, Nobel lecture: Sand from centuries past: Send future voices fast. Rev. Mod. Phys. **82**, 2299–2303 (2010). [Online]. Available: https://link.aps.org/doi/10.1103/RevModPhys.82. 2299
19. F.P. Kapron, D.B. Keck, R.D. Maurer, Radiation losses in glass optical waveguides. Appl. Phys. Lett. **17**(10), 423–425 (1970)
20. S.E. Miller, E.A.J. Marcatili, T. Li, Research toward optical-fiber transmission systems. Proc. IEEE **61**(12), 1703–1704 (1973)
21. T. Miya, Y. Terunuma, T. Hosaka, T. Miyashita, Ultimate low-loss single-mode fibre at 1.55 μm. Electronics Letters **15**(4), 106–108 (1979)
22. T. Li, Optical fiber communication - the state of the art. IEEE Trans. Commun. **26**(7), 946–955 (1978)
23. T.L. Maione, D.D. Sell, D.H. Wolaver, Atlanta fiber system experiment: practical 45-Mb/s regenerator for lightwave transmission. Bell Syst. Tech. J. **57**(6), 1837–1856 (1978)
24. T. Li, Advances in optical fiber communications: an historical perspective. IEEE J. Sel. Areas Commun. **1**(3), 356–372 (1983)
25. E. Iwahashi, Trends in long-wavelength single-mode transmission systems and demonstrations in Japan. IEEE J. Quantum Electron. **17**(6), 890–896 (1981)
26. K.R. Preston, K.C. Woollard, K.H. Cameron, External cavity controlled single longitudinal mode laser transmitter module. Electronics Letters **17**(24), 931–933 (1981)
27. P. Henry, Introduction to lightwave transmission. IEEE Commun. Mag. **23**(5), 12–16 (1985)
28. S.B. Poole, D.N. Payne, M.E. Fermann, Fabrication of low-loss optical fibres containing rare-earth ions. Electronics Letters **21**(17), 737–738 (1985)
29. R.J. Mears, L. Reekie, I.M. Jauncey, D.N. Payne, Low-noise erbium-doped fibre amplifier operating at 1.54μm. Electronics Letters **23**(19), 1026–1028 (1987)
30. E. Desurvire, J.R. Simpson, P.C. Becker, High-gain erbium-doped traveling-wave fiber amplifier. Optics Letters **12**(11), 888–890 (1987)
31. J. Cai, H.G. Batshon, M.V. Mazurczyk, O.V. Sinkin, D. Wang, M. Paskov, C.R. Davidson, W.W. Patterson, A. Turukhin, M.A. Bolshtyansky, D.G. Foursa, 51.5 Tb/s capacity over 17,107 km in C+L bandwidth using single-mode fibers and nonlinearity compensation. J. Lightwave Technol. **36**(11), 2135–2141 (2018)
32. P.J. Winzer, G.J. Foschini, MIMO capacities and outage probabilities in spatially multiplexed optical transport systems. Optics Express **19**(17), 16680–16696 (2011)
33. K.-P. Ho, J.M. Kahn, Mode-dependent loss and gain: statistics and effect on mode-division multiplexing. Optics Express **19**(17), 16612–16635 (2011)
34. D.M. Marom, P.D. Colbourne, A. D'Errico, N.K. Fontaine, Y. Ikuma, R. Proietti, L. Zong, J.M. Rivas-Moscoso, I. Tomkos, Survey of photonic switching architectures and technologies in support of spatially and spectrally flexible optical networking. J. Opt. Commun. Networking **9**(1), 1–26 (2017). [Online]. Available: http://jocn.osa.org/abstract.cfm?URI=jocn-9-1-1
35. K. Igarashi, T. Tsuritani, I. Morita, Y. Tsuchida, K. Maeda, M. Tadakuma, T. Saito, K. Watanabe, K. Imamura, R. Sugizaki, M. Suzuki, 1.03-exabit/s·km super-Nyquist-WDM transmission over 7,326-km seven-core fiber, in *Proceedings of European Conference on Optical Communication (ECOC)*, 2013, pp. 1–3
36. T. Kobayashi, H. Takara, A. Sano, T. Mizuno, H. Kawakami, K. Miyamoto, K. Hiraga, Y. Abe, H. Ono, M. Wada, Y. Sasaki, I. Ishida, K. Takenaga, S. Matsuo, K. Saitoh, M. Yamada, H. Masuda, T. Morioka, 2 x 344 Tb/s propagation-direction interleaved transmission over 1500-km MCF enhanced by multicarrier full electric-field digital back-propagation, in *Proceedings of European Conference on Optical Communication (ECOC)*, 2013, pp. 1–3
37. A. Turukhin, H.G. Batshon, M. Mazurczyk, Y. Sun, C.R. Davidson, J. Chai, O.V. Sinkin, W. Patterson, G. Wolter, M.A. Bolshtyansky, D.G. Foursa, A. Pilipetskii, Demonstration

of 0.52 Pb/s potential transmission capacity over 8,830 km using multicore fiber, in *42nd European Conference on Optical Communication (ECOC 2016)*, 2016, pp. 1–3

38. B.J. Puttnam, G. Rademacher, R.S. Luís, T.A. Eriksson, W. Klaus, Y. Awaji, N. Wada, K. Maeda, S. Takasaka, R. Sugizaki, 0.715 Pb/s transmission over 2,009.6 km in 19-core cladding pumped EDFA amplified MCF link, in *Proceedings of Optical Fiber Communication Conference (OFC)*, 2019, pp. 1–3

39. R. Ramaswami, K. Sivarajan, G. Sasaki, *Optical Networks: A Practical Perspective*, 3rd edn. (Morgan Kaufmann Publishers, San Francisco, CA, USA, 2009)

40. I. Chlamtac, A. Ganz, G. Karmi, Purely optical networks for terabit communication, in *IEEE INFOCOM '89, Proceedings of the Eighth Annual Joint Conference of the IEEE Computer and Communications Societies*, vol. 3, April 1989, pp. 887–896

41. O. Gerstel, M. Jinno, A. Lord, S.J.B. Yoo, Elastic optical networking: a new dawn for the optical layer? IEEE Commun. Mag. **50**(2), s12–s20 (2012)

42. A.L.N. Souza, E.J.M. Ruiz, J.D. Reis, L.H.H. Carvalho, J.R.F. Oliveira, D.S. Arantes, M.H.M. Costa, D.A.A. Mello, Parameter selection in optical networks with variable-code-rate superchannels. J. Opt. Commun. Networking **8**(7), A152–A161 (2016)

43. M. Jinno, H. Takara, B. Kozicki, Y. Tsukishima, Y. Sone, S. Matsuoka, Spectrum-efficient and scalable elastic optical path network: architecture, benefits, and enabling technologies. IEEE Commun. Mag. **47**(11), 66–73 (2009)

44. D.A.A. Mello, A.N. Barreto, T.C. de Lima, T.F. Portela, L. Beygi, J.M. Kahn, Optical networking with variable-code-rate transceivers. J. Lightwave Technol. **32**(2), 257–266 (2014). [Online]. Available: http://jlt.osa.org/abstract.cfm?URI=jlt-32-2-257

45. P.E. Green, Optical networking update. IEEE J. Sel. Areas Commun. **14**(5), 764–779 (1996)

46. T.A. Strasser, J.L. Wagener, Wavelength-selective switches for ROADM applications. IEEE J. Sel. Top. Quantum Electron. **16**(5), 1150–1157 (2010)

47. J.E. Ford, V.A. Aksyuk, D.J. Bishop, J.A. Walker, Wavelength add-drop switching using tilting micromirrors. J. Lightwave Technol. **17**(5), 904–911 (1999)

48. G. Baxter, S. Frisken, D. Abakoumov, Hao Zhou, I. Clarke, A. Bartos, and S. Poole, Highly programmable wavelength selective switch based on liquid crystal on silicon switching elements, in *Proceedings of Optical Fiber Communication Conference and National Fiber Optic Engineers Conference (OFC/NFOEC)*, March 2006

49. S. Gringeri, B. Basch, V. Shukla, R. Egorov, T.J. Xia, Flexible architectures for optical transport nodes and networks. IEEE Commun. Mag. **48**(7), 40–50 (2010)

50. R. Younce, J. Larikova, Y. Wang, Engineering 400G for colorless-directionless-contentionless architecture in metro/regional networks [invited]. J. Opt. Commun. Networking **5**(10), A267–A273 (2013)

Chapter 2
The Optical Transmitter

Coherent detection and digital signal processing (DSP) are now essential building blocks of modern optical communications. However, it was not always that way. As we have seen in Chap. 1, until the mid-2000s on–off keying (OOK) amplitude modulation with direct detection was practically the only modulation format used in commercial systems. Looking back in time, it took more than forty years for multilevel modulation formats to be widely deployed. The reason for this relatively late adoption was that, actually, bandwidth in wavelength division multiplexing (WDM) systems was abundant, and there was no pressure for improving spectral efficiency. As this pressure appeared in the mid-2000s, the optical communications community moved fast to develop technologies for spectrally efficient transmission. Nowadays, phase and amplitude modulation formats are extensively used in optical communication systems, especially quadrature amplitude modulation (QAM) formats [1, 2]. This chapter is dedicated to the main technologies that underpin modern optical transmitters. As current optical systems resort to several concepts from classical communication theory, we begin this chapter by reviewing the fundamentals of digital communications and later explore the details of an optical transmitter.

2.1 Principles of Digital Modulation

When studying communication theory, the first step is to understand how to represent signals in the different stages of the system. In the transmission by a communication channel, in our case an optical fiber, the intensity and phase of an approximately sinusoidal carrier are modified to convey information. This process is called modulation. The modulated signal can be represented as

$$x(t) = \sqrt{2}A(t)\cos(2\pi f_c t + \phi(t)), \qquad (2.1)$$

© Springer Nature Switzerland AG 2021
D. A. de Arruda Mello, F. A. Barbosa, *Digital Coherent Optical Systems*,
Optical Networks, https://doi.org/10.1007/978-3-030-66541-8_2

where f_c is the carrier frequency, and $A(t)$ and $\phi(t)$ are the amplitude and phase modulation components. The $\sqrt{2}$ factor normalizes the energy of the $\cos(\cdot)$ function. We call this a passband signal representation. A more convenient way to write (2.1) is [3]

$$x(t) = \sqrt{2}x_I(t)\cos(2\pi f_c t) - \sqrt{2}x_Q(t)\sin(2\pi f_c t), \qquad (2.2)$$

where $x_I(t) = A(t)\cos(\phi(t))$ and $x_Q(t) = A(t)\sin(\phi(t))$ are the in-phase and quadrature components. Here, we assume that $f_c > W/2$, where W is the spectral support of $x_Q(t)$ and $x_I(t)$.

It is possible to adopt another representation of the transmitted signals that suppress the dependence on the carrier frequency using complex numbers, called the baseband representation. The baseband representation of $x(t)$ is

$$x_b(t) = x_I(t) + ix_Q(t). \qquad (2.3)$$

Note that $x(t)$ can be easily obtained back from $x_b(t)$ by

$$x(t) = \sqrt{2}\mathbb{R}\{x_b(t)e^{i2\pi f_c t}\}. \qquad (2.4)$$

In digital transmission systems with passband pulse amplitude modulation (PAM), the continuous-time signals $x_I(t)$ and $x_Q(t)$ can be expressed as a sum of time-delayed continuous waveforms:

$$x_I(t) = \sum_k x_k^I g(t - kT_s), \qquad (2.5)$$

$$x_Q(t) = \sum_k x_k^Q g(t - kT_s), \qquad (2.6)$$

where T_s is the symbol period.

We assume that signals $x_I(t)$ and $x_Q(t)$ have the same pulse shape $g(t)$ and discrete in-phase and quadrature components x_k^I and x_k^Q. Depending on the choice of x_k^I and x_k^Q, various modulation formats, represented by their constellations, are implemented. Figure 2.1 shows common constellations used in digital communications. The OOK constellation (Fig. 2.1a) is the simplest and has been the most used throughout the history of optical communications, from its conception to the present day. It has only two symbols, one with amplitude 0 and the other with amplitude A. In intra-office applications, OOK modulation is still the most used. However, this scenario is changing quickly, and simplified coherent transceivers with multilevel modulation formats should become popular in short-reach applications as coherent transceivers evolve toward smaller footprints and power consumption [4].

The second simplest modulation format is binary phase-shift keying (BPSK), shown in Fig. 2.1b. In BPSK, the two transmitted constellation symbols are in symmetric position with respect to the origin. At reception, coherent detection is

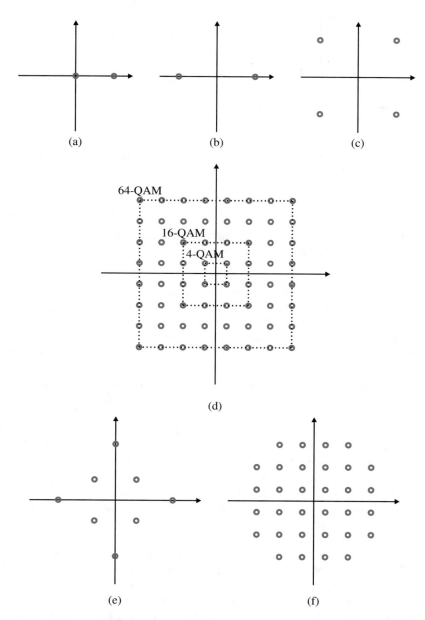

Fig. 2.1 Typical symbol constellations used in digital communications. (**a**) OOK. (**b**) BPSK. (**c**) QPSK. (**d**) Square M-QAM. (**e**) Star 8-QAM. (**f**) 32-QAM

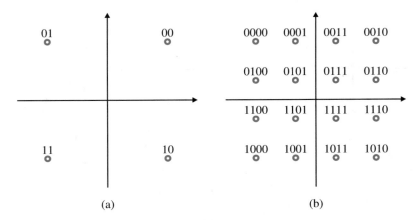

Fig. 2.2 Gray mapping for (**a**) QPSK; (**b**) 16-QAM. Gray mapping ensures that neighboring symbols differ by only one bit

required for polarity discrimination. Differential detection, with modulation formats such as differential phase-shift keying (DPSK), has also been briefly investigated in optical communications [5] but were replaced by coherent detection because of advantages such as seamless chromatic dispersion and polarization mode dispersion compensation. BPSK uses only one degree of freedom, i.e., the modulation of the in-phase component. The simplest modulation format that exploits the two degrees of freedom provided by the complex plane is the quadrature phase-shift keying (QPSK) format, whose constellation is shown in Fig. 2.1c. The QPSK constellation consists of the sum of two BPSK constellations in quadrature. QPSK and BPSK are considered phase modulation formats because their amplitude is kept constant for all the constellation symbols. A widely used family of modulation formats that achieves excellent performance is the M-ary quadrature amplitude modulation (M-QAM). In M-QAM, $\sqrt{M}$ amplitude levels of the in-phase and quadrature components are uniformly distributed and independently modulated. Usually, $M = L^2$, and L is a power of two. Figure 2.1d shows M-QAM modulation formats, for $M = 4$, 16, and 64. Other non-square M-QAM constellations can also be defined, such as star 8-QAM (Fig. 2.1e) and 32-QAM (Fig. 2.1f).

Another important issue is how to map bits into constellation symbols. It is desirable that adjacent symbols are mapped to code words that differ by a small number of bits, ideally only one. In this way, even if one symbol conveys $m = \log_2(M)$ bits, most symbol errors cause only one bit error. This is achieved by an approach known as Gray mapping, as presented in Fig. 2.2 for the QPSK and 16-QAM modulation formats.

There are several metrics to evaluate the performance of a modulation format, one of them is spectral efficiency (SE). Spectral efficiency itself can be defined in several ways, but we define it as the ratio between the conveyed net bit rate, R_b, and the bandwidth required to support this rate, W,

$$\text{SE} = \frac{R_b}{W}. \tag{2.7}$$

The spectral support W of the channel depends directly on the pulse shape $g(t)$. The Nyquist criterion determines the existence of a family of pulse shapes $g(t)$ for which transmission is carried out free of intersymbol interference (ISI) [6]. This condition for ISI-free transmission can be expressed as [7]

$$\frac{1}{T_s} \sum_{m=-\infty}^{\infty} G\left(f - \frac{m}{T_s}\right) = 1, \tag{2.8}$$

where $G(f)$ is the Fourier transform of $g(t)$. ISI-free transmission is achieved if the sum of frequency-shifted versions of $G(f)$ add up to a constant. Pulses satisfying this condition are called *Nyquist pulses*. The Nyquist pulse shape with minimum transmission bandwidth, $1/(2T_s)$, is obtained by a sinc pulse shape

$$g(t) = \frac{\sin(\pi t/T_s)}{\pi t/T_s}. \tag{2.9}$$

Pulses based on the sinc function have infinite duration, which hinders their practical application. Alternatively, raised cosine (RC) pulses are Nyquist pulses whose spectral duration and support are controlled by a roll-off factor β^{RC} ($0 \leq \beta^{\text{RC}} \leq 1$). The frequency response of the RC shaping filter in given by

$$H_{\text{RC}}(f) = \begin{cases} 1, & \text{if } |f| < \frac{1-\beta^{\text{RC}}}{2T_s}; \\ 0, & \text{if } |f| > \frac{1+\beta^{\text{RC}}}{2T_s}; \\ 0.5 + 0.5\cos\left(\frac{\pi T_s}{\beta^{\text{RC}}}\left(|f| - \frac{1-\beta^{\text{RC}}}{2T_s}\right)\right), & \text{if } \frac{1+\beta^{\text{RC}}}{2T_s} > |f| > \frac{1-\beta^{\text{RC}}}{2T_s}, \end{cases} \tag{2.10}$$

The transfer function $H_{RC}(f)$, as well as the corresponding impulse response, $h_{RC}(t)$, are shown in Fig. 2.3. For $\beta^{\text{RC}} = 0$, the signal has a rectangular frequency response and an infinite duration in time. By increasing β^{RC}, the frequency response widens, but the pulse becomes more time-constrained. Coherent optical systems can work with fairly low roll-off factors, ranging from 0.01 to 0.1, implementing a pulse with near-rectangular spectrum. However, the generation of pulses with such low roll-off factors is complex, requiring shaping filters with hundreds of coefficients.

Once generated, the signal passes through a channel and is retrieved at the receiver. The most common channel model is the additive white Gaussian noise (AWGN) channel, for which the received signal $r(t)$ is given by

$$r(t) = x(t) + \eta(t), \tag{2.11}$$

(a)

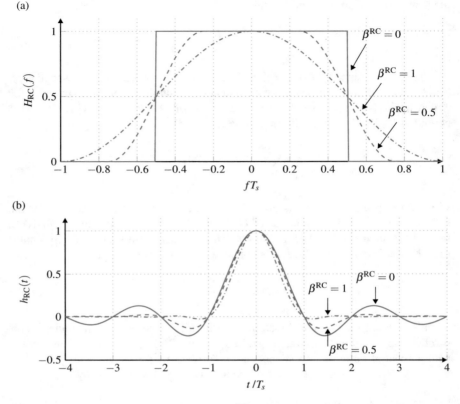

(b)

Fig. 2.3 Raised cosine (RC) filter for different β^{RC} and T_s. (**a**) Frequency response. (**b**) Impulse response

where $\eta(t)$ is the white Gaussian noise with a power spectral density equal to $N_0/2$ over the entire spectrum. The noise process $\eta(t)$ has infinite power, being therefore a fictitious process. Anyway, it is useful under the consideration that $r(t)$ is filtered at the receiver, and the noise power of interest is that within the signal bandwidth.

Considering the AWGN channel, the optimum receiver according to the minimum distance criterion is composed of two stages [3]. Assuming complex baseband representation, the first stage is a filter applied to the received signal $r_b(t)$, as shown in Fig. 2.4. This filter is matched to the transmitted waveform, having an impulse response given by $g^*(-t)$. The second stage is a sampler generating T_s-spaced samples at optimum sampling instants of maximum energy.

Because of the multiplication of the signal Fourier transform by the corresponding matched filter frequency response, a combined raised cosine shape can be obtained by a root-raised cosine (RRC) shaping filter at the transmitter

$$r_b(t) \longrightarrow \boxed{g^*(-t)} \longrightarrow \overset{\times}{\underset{kT_s}{\bullet}} \bullet \longrightarrow y_k = x_k + n_k$$

Fig. 2.4 Optimum receiver according to the minimum distance criterion. The receiver filter impulse response $g^*(-t)$ is matched to the transmitted pulse shape $g(t)$. The signal is sampled with rate $1/T_s$ at the instants of maximal energy. Without loss of generality, assuming $\int_{-\infty}^{-\infty} |g(t)|^2 dt = 1$, the output is given by the transmitted sample x_k added to a noise sample n_k with variance N_0 ($N_0/2$ for the in-phase and quadrature components)

$$g(t) = h_{\text{RRC}}(t) = \mathbb{F}^{-1}\left(\sqrt{H_{\text{RC}}(f)}\right). \tag{2.12}$$

Without loss of generality, we assume $\int_{-\infty}^{+\infty} |g(t)|^2 dt = 1$. Then, after the matched filter, the discrete complex baseband signal y_k is the sum of the original signal components x_k with a white Gaussian noise component η_k [8]

$$y_k = x_k + \eta_k. \tag{2.13}$$

The in-phase and quadrature components of y_k are given by

$$y_k^I = x_k^I + \eta_k^I, \tag{2.14}$$

$$y_k^Q = x_k^Q + \eta_k^Q, \tag{2.15}$$

where η_k^I and η_k^Q are zero-mean white Gaussian noise processes with variance $N_0/2$ in the in-phase and quadrature components. The signal-to-noise ratio (SNR), calculated as the ratio of signal and noise powers considering both the in-phase and quadrature components, can be expressed as

$$\text{SNR} = \frac{E_s}{N_0},$$

where $E_s = E[|x_k|^2]$.

This definition does not take into account the amount of information bits per symbol conveyed by the transmitted signal. Therefore, it is usual to evaluate an alternative SNR defined in terms of the signal power per bit, as

$$\text{SNR}_b = \frac{E_b}{N_0},$$

where E_b is the power per bit ($E_b = E_s/b$), and b is the number of conveyed bits per symbol.

The transmission of M-QAM signals over an AWGN channel results in a symbol error probability (SER) given by [8]

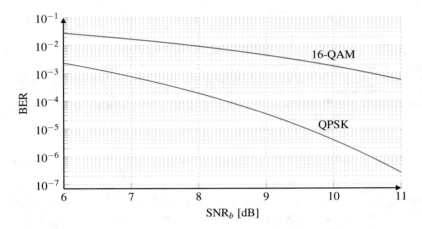

Fig. 2.5 Bit error rate (BER) of the QPSK and 16-QAM formats over an AWGN channel, assuming Gray mapping

$$p_e \approx 2 \left(1 - \frac{1}{\sqrt{M}}\right) \mathrm{erfc}\left(\sqrt{\frac{3E_s}{2(M-1)N_0}}\right). \tag{2.16}$$

For example, for the QPSK modulation format, (2.16) reduces to

$$p_e \approx \mathrm{erfc}\left(\sqrt{\frac{E_s}{2N_0}}\right). \tag{2.17}$$

For Gray mapping, a symbol error results in usually only one bit error. Therefore, the bit error probability is given by

$$p_b \approx \frac{1}{b} p_e. \tag{2.18}$$

Optical transmission systems are usually evaluated in terms of the bit error rate (BER), calculated as the ratio between the number of bit errors and the number of transmitted bits. The BER approaches the bit error probability for sufficiently long observation windows. Figure 2.5 shows the BER for the 16-QAM and QPSK modulation formats. Current optical communication systems resort to last-generation forward error correction (FEC) schemes. Usually, pre-FEC BERs below 10^{-3} to 10^{-2} are needed to achieve a post-FEC BER below 10^{-15}, required in modern optical communication systems. A more in-depth discussion about the use of pre-FEC BERs as a performance metric in optical communication systems is presented in Chap. 8.

2.2 Lasers

In the last section, we discussed principles of digital modulation. The modulation process modifies certain properties of a sinusoidal carrier to convey information. In particular, we discussed the QAM modulation format, which modifies the carrier amplitude and phase. But we gave no further details on how this carrier is generated. In optical communication systems, the carrier is usually produced by semiconductor lasers and light-emitting diodes (LEDs). For long-distance transmission with coherent detection, which is the focus of this book, the use of LEDs is prohibitive because of their large bandwidths (usually called linewidth in the scope of oscillators). Performance losses would arise mainly because it becomes very difficult to track the phase at the receiver. Therefore, we assume throughout this book that all optical carriers are generated by lasers. The name laser appeared initially as an acronym for *light amplification by stimulated emission of radiation*. Since its invention in the late 1950s [9, 10], lasers have been applied in the most diverse industries, from medicine to manufacturing and telecommunications. The main advantages of lasers over other light sources are their collimated beam, relatively high intensities and coherence. For long-distance optical transmission, coherence is of primary importance, particularly for systems with phase modulation and coherent detection. As we modulate the phase of the generated carrier, phase noise must be relatively low to avoid impairments in the detection process. Such distinctive properties are achieved by two physical processes, stimulated emission [11] and cavity resonance. Figure 2.6 depicts the process of stimulated emission in a simplified two-level system, along with the absorption and spontaneous emission processes [12]. In absorption, one photon is absorbed, triggering an atom transition from ground state to excited state. In spontaneous emission, one photon is emitted, while an atom migrates from the excited state to the ground state. In stimulated emission, one propagating photon triggers the generation of a second photon, with similar characteristics (frequency, polarization, and direction of propagation) as the original one [13]. In thermal equilibrium, there are more atoms in lower energy states than in excited energy states, and absorption exceeds stimulated emission. Lasers, however, require the opposite regime, where stimulated emission dominates over absorption, effectively amplifying the input signal. This is achieved by the so-called population inversion, through which atoms are pumped from the ground state to the excited state. In semiconductor lasers, pumping is carried out electrically taking into account the bands of the semiconductor device.

As we mentioned before, the next requirement for a laser is a cavity. The cavity with gain medium enabled by stimulated emission oscillates in a set of resonant frequencies called longitudinal modes. The intensity of a specific longitudinal mode depends on the relationship between gain and loss in the operating frequency. Lasing modes will be those for which the gains in the cavity outweigh the losses. Figure 2.7a illustrates the gain and loss profile of a semiconductor laser, and Fig. 2.7b depicts the corresponding spectral shape. Semiconductor lasers are typically a p–n junction with a thin active layer of a different material between

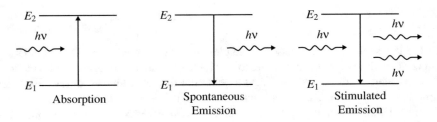

Fig. 2.6 Absorption, spontaneous emission, and stimulated emission processes in lasers. In absorption, one photon is absorbed, triggering an atom transition from ground state to excited state. In spontaneous emission, one photon is emitted, while an atom migrates from the excited state to the ground state. In stimulated emission, one propagating photon triggers the generation of a second photon, with similar characteristics (frequency, polarization, and direction of propagation) as the original one

the p- and n-type layers. The cavity is obtained by the cleaved facets of the device that work as mirrors. Light is collected laterally, as shown in Fig. 2.8. The active layer role is twofold. First, it confines carriers within an active area, enhancing optical amplification. Second, its refractive index is higher than that in the cladding layers, behaving therefore as a planar waveguide. Population inversion in the semiconductor is obtained by the electric pumping of carriers, generated by forward-biasing the junction.

Lasers used in coherent detection require a particularly narrow linewidth, in order to serve as a suitable reference source for phase detection. As shown in Fig. 2.7b, the various longitudinal modes of a semiconductor laser, and its wide spectrum, would make long-distance transmission prohibitive. For this reason, single longitudinal-mode lasers (SLMs) are used in these applications. In fact, the other longitudinal modes are suppressed, maintaining only the desired operating mode. This is achieved in different ways, as shown in Fig. 2.9a. In distributed-feedback lasers (DFB), the undesired longitudinal modes are suppressed by placing a Bragg grating within the laser cavity [14]. Current coherent optical systems use mostly DFB lasers with linewidths of some hundreds of kHz. In external-cavity lasers (ECLs), an external filter is used to filter out longitudinal modes. Figure 2.9b shows an architecture with an external grating that reflects back into the waveguide the desired operating frequency and reflects out of the waveguide the undesired frequencies. The output power is collected by the zero-order reflection of the grating.

Assuming that the laser output has constant intensity (thus neglecting intensity noise), the electric field in its output is given by

$$E_l(t) = E_0 e^{j\phi_n(t)}, \tag{2.19}$$

where E_0 is a constant field amplitude and $\phi_n(t)$ is the phase noise process. The non-zero linewidth of lasers gives rise to a phase noise process that is well-modeled by a continuous-time Wiener process, such that the phase difference between two

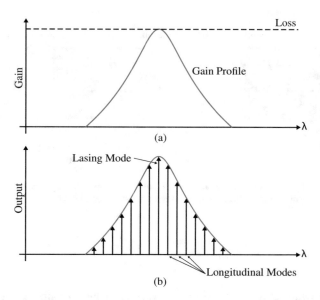

(a)

(b)

Fig. 2.7 (**a**) Loss and gain profile of a semiconductor laser. The lasing mode requires the cavity losses to be compensated by the cavity gain. (**b**) Longitudinal modes of the laser. The intensity of a longitudinal mode varies according to the balance of gain and loss in the cavity

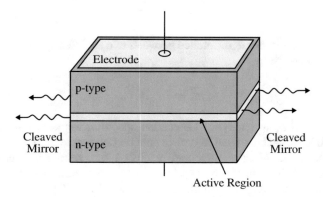

Fig. 2.8 Architecture of a semiconductor laser. An active layer is surrounded by p-type and n-type cladding layers. The p–n junction is forward biased, and carriers are recombined in the active region, generating photons in the recombination process. The cleaved facets of the structure work as the mirrors of a cavity

samples collected τ seconds apart, $\phi_n(t) - \phi_n(t - \tau)$, is described by a zero-mean Gaussian process with variance $\sigma_{n_\phi}^2 |\tau|$, where $\sigma_{n_\phi}^2$ is the variance of the instantaneous frequency [5]. As a consequence, the power spectrum density of E_l, S_l, acquires a Lorentzian shape

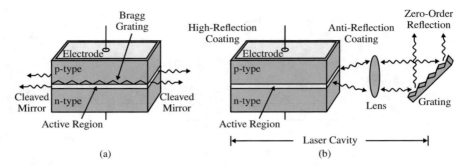

Fig. 2.9 (**a**) DFB laser, longitudinal modes are suppressed by a Bragg grating along the active region. (**b**) ECL, longitudinal modes are suppressed by an external cavity

$$S_l(f) = \frac{\sigma_{n_\phi}^2}{\frac{1}{4}\sigma_{n_\phi}^4 + (2\pi f)^2}. \tag{2.20}$$

The full width at a half maximum (FWHM) of the Lorentzian spectrum, also known as the laser linewidth, is given by [5]

$$\Delta v = \frac{\sigma_{n_\phi}^2}{2\pi}. \tag{2.21}$$

Thus, the variance in phase difference between two samples spaced by τ samples, σ_ϕ^2, can also be given by

$$\sigma_\phi^2 = 2\pi \Delta v |\tau|. \tag{2.22}$$

Figure 2.10 shows the (analytic) Lorentzian spectrum for laser with a linewidth equal to 100 kHz, as those used in typical coherent optical systems.

The phase noise of transmitter and receiver lasers results in rotations of the received constellations. In coherent optical systems, these rotations are compensated in DSP by phase recovery algorithms (for more details, see Chap. 6).

2.3   Optical Modulation

In the last section, we learned some basic concepts of digital modulation, and how information is conveyed by modifying the amplitude and phase of a carrier. We also learned that semiconductor lasers can generate carriers with relatively narrow linewidth, with some tens or some few hundreds of kHz. Optical transmitters follow the same principle as any passband communication system. First, bits of information

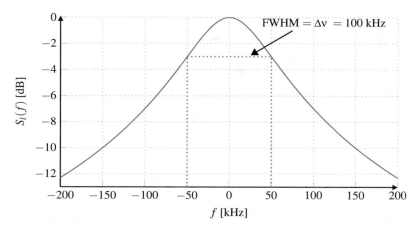

Fig. 2.10 Lorentzian spectrum with 100 kHz linewidth. In the figure, $f = 0$ indicates the laser center frequency

are sent from a data source to a symbol mapper that performs the assignment of symbols according to the constellation of a specific modulation format. Then, a shaping filter is used to generate pulses conveying the mapped symbols. This is all carried out in the digital domain, i.e., in a digital signal processor. Therefore, before transmitting the signals over the channel, digital-to-analog converters (DACs) are required to generate analog waveforms. Finally, these signals are sent to an optical modulator that converts the input analog electric signals into a modulated optical carrier.

In fiber optic communication systems, direct modulation is the simplest technique used to modulate an optical carrier. Direct modulation modifies the driving current of a semiconductor laser causing variations in the amplitude of the generated optical signal according to the information to be sent. As a disadvantage, direct modulation produces pulses that are affected by *chirp*, which are continuous variations of the optical frequency during a symbol period [12, 15]. Chirp causes spectral broadening and limits the transmission rate in several ways, for example, impairing phase modulation. A further disadvantage of direct modulation is the inability to follow fast variations in the driving voltage. In spite of these disadvantages, direct modulation is widely used in short-range optical systems with intensity modulation and direct detection, mainly because of its simplicity and low cost.

Alternatively, external modulation is used in most high-rate systems. This modulation technique requires the use of an extra component, the optical modulator, which is coupled at the output of the optical source. Variations of phase, amplitude, or a combination of the two can be performed according to electric signals representing the information to be sent. Although increasing transmitter complexity, external modulation provides a considerably better system performance compared with direct modulation by minimizing several undesired effects, particularly chirp and bandwidth limitations. Recent external modulator architectures also allow to

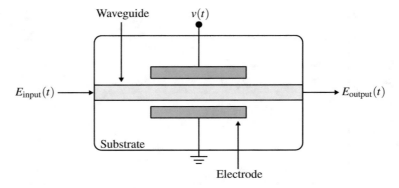

Fig. 2.11 Phase modulator architecture. The input signal is guided through a planar dielectric waveguide. The applied voltage modulates the phase of the guided signal using the electro-optical effect

exploit the polarization dimension, doubling the capacity of the optical channel. The main building block of these advanced modulators is the phase modulator presented below.

2.3.1 Phase Modulator

The phase modulator (PM) is a device that relies on the Pockels effect to imprint variations in the phase of an optical carrier [16]. Such a device is comprised of a waveguide, which is generally lithium niobate (LiNbO3), surrounded by a pair of electrodes that enable the application of an electric voltage. The basic scheme of a phase modulator is shown in Fig. 2.11, where $E_{input}(t)$ and $E_{output}(t)$ are the electric fields of the input and output optical signals, respectively, and $v(t)$ is the electric signal that drives the PM.

According to the Pockels effect, also known as electro-optical effect, the application of an external electric voltage in certain materials is able to modify their refractive index. Consequently, the effective refractive index of a waveguide [17] made of these materials can also be modified. Using this process, a modulated driving voltage can also modulate the phase of the optical carrier traversing the waveguide according to the information to be transmitted. The phase displacement φ_{PM} is given by [17]

$$\varphi_{PM}(t) = \frac{2\pi}{\lambda} \Delta n_{eff}(t) l_i, \tag{2.23}$$

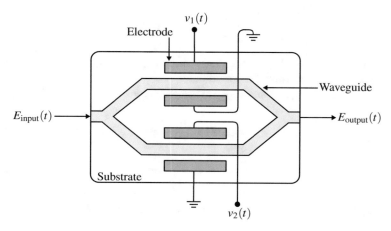

Fig. 2.12 Mach–Zehnder modulator architecture. The outputs of two phase modulators are combined to produce constructive (maximum amplitude) or destructive (minimum amplitude) interference

where λ is the operating wavelength, $\Delta n_{eff}(t)$ is the variation of the effective refractive index, and l_i is the interaction length. In general, $\Delta n_{eff}(t)$ is linearly proportional to $v(t)$, and the proportionality constant is usually presented in terms of the voltage that generates a phase displacement of π rad, known as V_π,

$$\frac{E_{output}(t)}{E_{input}(t)} = e^{j\varphi_{PM}(t)} = e^{j\frac{\pi}{V_\pi}v(t)}. \tag{2.24}$$

2.3.2 Mach–Zehnder Modulator

The Mach–Zehnder modulator (MZM) is a device that uses the principle of interference between propagating signals to generate amplitude and phase modulation. Its name stems from the fact that the structure employed to generate interference between the propagating signals is based on a Mach–Zehnder interferometer (MZI), as illustrated in Fig. 2.12. The optical signal entering the structure is divided into two distinct arms. In each of these arms, a phase modulator acts on the propagating optical signal by performing a phase shift proportional to the applied electric voltage. After this process, the optical signals from the two arms are recombined, and depending on the phase difference between them, a constructive or destructive interference pattern is observed. This device structure gives rise to the transfer function

$$\frac{E_{\text{output}}(t)}{E_{\text{input}}(t)} = \frac{1}{2}\left(e^{j\varphi_{PM_1}(t)} + e^{j\varphi_{PM_2}(t)}\right), \tag{2.25}$$

where $\varphi_{PM_1}(t)$ and $\varphi_{PM_2}(t)$ are the phase shifts experienced by the optical signals in the upper and lower arms of the modulator.

The phase shifts $\varphi_{PM_1}(t)$ and $\varphi_{PM_2}(t)$ are given by

$$\varphi_{PM_1}(t) = \frac{v_1(t)}{V_{\pi_1}}\pi, \tag{2.26}$$

$$\varphi_{PM_2}(t) = \frac{v_2(t)}{V_{\pi_2}}\pi, \tag{2.27}$$

where V_{π_1} and V_{π_2} are the V_π parameters for the phase modulator in each of the arms.

The MZM operation mode depends on the relationship between the electric voltages that drive the phase modulators. There are two main operation modes. If $\varphi_{PM_1}(t) = \varphi_{PM_2}(t) = \varphi(t)$, the MZM is said to operate in *push–push* mode [17]. In this case, the interference between the optical signals is always constructive, and only pure phase modulation is achieved. If $\varphi_{PM_1}(t) = -\varphi_{PM_2}(t) = \varphi(t)$, the MZM is said to operate in *push–pull* mode. The push–pull configuration allows to drive the MZM as an amplitude modulator. One important feature of this operation mode is that amplitude modulation is carried out without any phase variation in the transition between constellation points, avoiding chirp [1].

In push–pull configuration, it is possible to rewrite (2.25) as

$$\frac{E_{\text{output}(t)}}{E_{\text{input}}(t)} = \frac{1}{2}\left(e^{j\varphi_{PM_1}(t)} + e^{-j\varphi_{PM_1}(t)}\right) = \cos[\varphi_{PM_1}(t)]. \tag{2.28}$$

Assuming that $v_1(t) = -v_2(t) = v(t)/2$, in a scenario where $V_{\pi_1} = V_{\pi_2} = V_\pi$, the field transfer function of a MZM in push–pull configuration is given by [17]

$$\frac{E_{\text{output}}(t)}{E_{\text{input}}(t)} = \cos\left[\frac{v(t)}{2V_\pi}\pi\right]. \tag{2.29}$$

The relationship between the MZM transfer function and the driving voltage $v(t)$ is shown in Fig. 2.13a. For $v(t) = 0$, the field amplitude is maximum. For $v(t) = -V_\pi$, the amplitude falls to zero. For $v(t) = -2V_\pi$, the amplitude is maximum again, but the field polarity inverts. As we will see later, the ability to modulate amplitude and polarity is essential in generating modulation formats represented by the in-phase and quadrature components. A direct current (DC) bias voltage b can be applied to the arms of the MZM configured in push–pull mode to generate a certain modulation format, such that $v(t) = b + m(t)$, where $m(t)$ corresponds to the constellation symbols. The most common bias points are the quadrature and minimum transmittance points [17]. The quadrature operation point

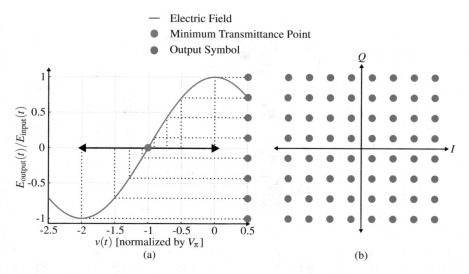

Fig. 2.13 (a) Transfer function of a MZM biased at the minimum transmittance point for the generation of an 8-PAM signal. (b) 64-QAM modulation format generated by the combination of two 8-PAM signals in quadrature

is achieved when the DC voltage applied to the MZM is equal to $b = -V_\pi/2 \pm kV_\pi$, where k is an integer. The minimum transmittance operation point is achieved for a bias voltage of $b = -V_\pi \pm 2kV_\pi$. In general, the quadrature point is used to generate amplitude modulation, while the minimum transmittance point is used to carry out amplitude and phase modulation.

2.3.3 In-Phase and Quadrature Modulator

The in-phase and quadrature modulator (IQM) is a component that is capable of generating constellation symbols at any position in the complex plane. To do so, its structure consists of a combination of two MZMs and one $\pi/2$ PM, as illustrated in Fig. 2.14. The optical signal entering the modulator is equally divided into two paths, one responsible for the in-phase component and the other for the quadrature component. In each path, a MZM, which is configured to operate in push–pull mode and at the point of minimum transmittance, modulates the amplitude and the polarity of the optical signal. The phase modulator positioned after the quadrature MZM is responsible for performing a rotation of $\pi/2$ rad on the modulated signal. Finally, the recombination of signals from both paths results in an optical symbol corresponding to the desired constellation point.

The modulation performed by the in-phase and quadrature MZMs causes the signals in each of the arms to experience a phase shift that can again be defined as a

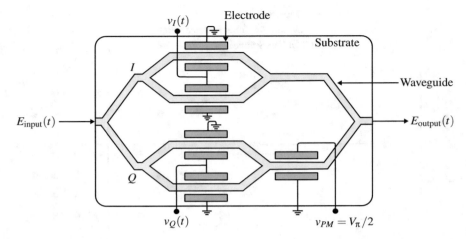

Fig. 2.14 In-phase and quadrature modulator (IQM). The IQM is constructed by two nested Mach–Zehnder modulators and one phase modulator responsible for generating a $\pi/2$ phase mismatch between the outputs. The v_I and v_Q voltages are modulated to generate the desired in-phase and quadrature components. Voltage v_{PM} is controlled to maintain the $\pi/2$ phase mismatch between the I and Q outputs

function of V_π (given for an MZM)

$$\varphi_I(t) = \frac{v_I(t)}{V_\pi}\pi, \tag{2.30}$$

$$\varphi_Q(t) = \frac{v_Q(t)}{V_\pi}\pi, \tag{2.31}$$

where $v_I(t)$ and $v_Q(t)$ are the voltages applied to the in-phase and quadrature MZMs. Thus, the transfer function of the electric field of an optical signal modulated by an IQM is given by [17]

$$\frac{E_{output}(t)}{E_{input}(t)} = \frac{1}{2}\cos\left[\frac{\varphi_I(t)}{2}\right] + j\frac{1}{2}\cos\left[\frac{\varphi_Q(t)}{2}\right]. \tag{2.32}$$

For a continuous wave (CW) input, $E_{input} = 1$ and E_{output} can take the form of any constellation point dictated by components $v_I(t)$ and $v_Q(t)$. The generation of a pulse amplitude modulation (PAM) signal for the in-phase or quadrature components of a 64-QAM constellation is shown in Fig. 2.13a. In the figure, the green circle indicates the bias on the point of minimum transmittance. A DC level of $-V_\pi$ is added to the input modulation voltage generated by the DAC. The excursion of the input signal then takes place from $-2V_\pi$ and 0. The example shows the generation of four output electric field intensity levels, and two polarities, making up an 8-ary pulse amplitude modulation (8-PAM) constellation with double polarity.

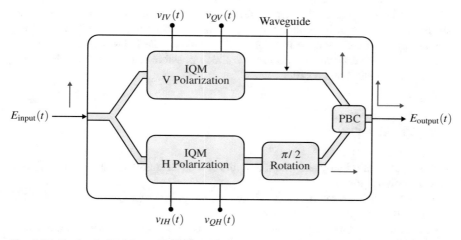

Fig. 2.15 Dual-polarization modulator (DPM). The DPM is generated by the combination of two IQMs and one $\pi/2$ polarization rotation. Each DPM is fed by two inputs, corresponding to the in-phase and quadrature components of a complex constellation

Using two of these MZMs, a 64-QAM constellation is generated, as shown in Fig. 2.13b.

2.3.4 Dual-Polarization Modulator

In addition to conveying information in the phase and amplitude of the optical signal, digital coherent optical systems also use polarization as an additional degree of freedom. Single-mode optical fibers support two degenerate (having the same propagation constant) optical modes, with orthogonal polarization orientations. Polarization multiplexing is a technique that allows two independent signals to be simultaneously transmitted in the orthogonal polarization orientations of an optical carrier, thus doubling the spectral efficiency of the system. The signals multiplexed in the two polarization orientations are mixed during propagation but still can be separated at the receiver by signal processing techniques. Figure 2.15 shows the basic architecture of a dual-polarization modulator (DPM) consisting of two IQMs, a $\pi/2$ rad polarization rotator and a polarization beam combiner (PBC).

The procedure of generating a polarization-multiplexed signal using a DPM starts with the balanced division of a CW optical carrier entering its structure in two paths. These paths are related to the vertical (V) and horizontal (H) polarization orientations that give rise to the polarized-multiplexed signal. In each of the two paths of the DPM, there is an IQM that modulates the optical signal according to the electric signals applied to it. In the V-path, the electric signals controlling the IQM are $v_{IV}(t)$ and $v_{QV}(t)$. In the H-path, the electric signals are $v_{IH}(t)$ and $v_{QH}(t)$.

The modulated signal of the H-path undergoes a polarization rotation of $\pi/2$ rad, so that it becomes orthogonal to the signal of the other path. Finally, the signals from both paths are combined in the PBC, producing the polarization-multiplexed signal. Alternative DPM architectures are also possible, e.g., using a polarization beam splitter (PBS) and a PBC.

2.4   Problems

1. Calculate analytically and plot the points of a QPSK and a 16-QAM constellation with unitary power.
2. Calculate the minimum distance between two points of a QPSK and a 16-QAM constellation with unitary power.
3. Calculate the relationship between SER and BER of a QPSK constellation, for all possible mappings. Note that mappings of a QPSK constellation are insensitive to rotations. Assume that errors only occur between the closest neighbors. Highlight the results obtained when Gray mapping is used.
4. Plot the normalized power spectrum of a laser with a 100-kHz linewidth and with a 50-kHz linewidth.
5. Simulate the transmission of a signal over an AWGN channel:

    (a) Generate a bit sequence and encode it using Gray mapping into a QPSK and a 16-QAM constellation. Normalize the generated symbol sequence to have unitary power. Plot the generated QPSK and 16-QAM constellations.
    (b) Generate the pulse-shaped waveform by convolving the generated symbol sequence with an RRC filter impulse response. Use an RRC filter with a roll-off factor $\beta^{RC} = 0.1$ and a span of 20 symbols, assuming 16 Sa/Symbol. Plot the normalized discrete impulse response of the RRC filter used for pulse shaping.
    (c) Add zero-mean, complex Gaussian AWGN (with variance $N_0/2$ in both the in-phase and quadrature components) to the pulse-shaped signal, yielding a given SNR_b. Note that it is necessary to consider the oversampling factor to properly scale the noise power according to the signal power per symbol. Pass the received signal through a filter matched to the transmitted waveform. This can be done using a code similar to the one used in item (b). Remember that it is necessary to downsample the received signal. Plot the resulting constellation after the matched filter.
    (d) Include a decision process to the simulator. Remember to use the same mapping scheme implemented at the transmitter. Reproduce Fig. 2.5. Compare the simulation with the theoretical curves. Use the same number of transmitted bits for both modulation formats.
6. Generate Nyquist-shaped 16-QAM electric signals and simulate the modulation of an optical carrier using an IQM. Use an RRC pulse shaping filter with a roll-off factor $\beta^{RC} = 0.1$, and a span of 20 symbols, assuming 16 Sa/Symbol.

For an input laser power of 0 dBm, plot the generated optical constellations (after downsampling the waveform to 1 sample per symbol and normalizing the downsampled signal to unitary power). Calculate the output optical power, for the following configurations:

Configuration	Bias b	Excursion $\min(m(t))$ to $\max(m(t))$
1	$-V_\pi$	$-2.3V_\pi$ to $0.3\ V_\pi$
2	$-V_\pi$	$-2V_\pi$ to $0\ V_\pi$
3	$-V_\pi$	$-1.5V_\pi$ to $-0.5V_\pi$
4	$-V_\pi$	$-1.2V_\pi$ to $-0.8V_\pi$
5	$-0.7V_\pi$	$-1.5V_\pi$ to $-0.5V_\pi$

Which is the best configuration? Justify.

2.5 Matlab/Octave Functions

2.5.1 Functions for Sect. 2.1

Matlab/Octave Code 2.1 QPSK and 16-QAM symbol generation

```
function [Bits,x] = SymbolGeneration(ModFormat,NSymb)
%%%%%%%%%%%%%%%%%%%%%%%%%%%%%%%%%%%%%%%%%%%%%%%%%%%%%%%%%%%%%%%%%%%%%%%%%%%%
% SYMBOLGENERATION [Bits,x] = SymbolGeneration(ModFormat,NSymb)          %
%                                                                        %
%   This function generates a sequence of bits 'Bits' with uniform       %
% distribution and, considering Gray mapping, generates a sequence of    %
% symbols 'x' according to the modulation format 'ModFormat'. The length %
% of 'x' is given by 'NSymb'. The symbols of 'x' follow the unitary power%
% constellation associated with 'ModFormat'. For transmission with pol.  %
% multiplexing, this function must be called twice.                      %
%                                                                        %
% Input:                                                                 %
%   ModFormat  = Modulation format: 'QPSK' or '16QAM';                   %
%   NSymb      = Number of symbols to be transmitted;                    %
%                                                                        %
% Output:                                                                %
%   Bits = Sequence of bits (column vector);                            %
%   x    = Sequence of symbols (column vector);                         %
%                                                                        %
% This function is part of the book Digital Coherent Optical Systems;    %
% Darli A. A. Mello and Fabio A. Barbosa;                               %
%%%%%%%%%%%%%%%%%%%%%%%%%%%%%%%%%%%%%%%%%%%%%%%%%%%%%%%%%%%%%%%%%%%%%%%%%%%%
    switch ModFormat
        case {'QPSK'}
            ModBits = 2;
        case '16QAM'
            ModBits = 4;
        otherwise
            error('The supported modulation formats are QPSK and 16-QAM;');
    end
```

```
% Generating the sequence of bits:
NBits = NSymb*ModBits; Bits = randi([0,1],NBits,1);

switch ModFormat
    case 'QPSK'
        % In-Phase and quadrature bits:
        BitsI = Bits(2:ModBits:NBits); BitsQ = Bits(1:ModBits:NBits);

        % QPSK modulation:
        xI = 1-2*BitsI; xQ = 1-2*BitsQ;

        % Normalized symbols:
        x = (1/sqrt(2))*(xI + xQ*1i);
    case '16QAM'
        % In-Phase and quadrature bits:
        BitsI1 = Bits(4:ModBits:NBits); BitsQ1 = Bits(3:ModBits:NBits);
        BitsI2 = Bits(2:ModBits:NBits); BitsQ2 = Bits(1:ModBits:NBits);

        % 16QAM modulation:
        % In-Phase:
        xI = ((~BitsI2 & ~BitsI1)*(+3) + ( BitsI2 & ~BitsI1)*(+1) + ...
               ( BitsI2 &  BitsI1)*(-1) + (~BitsI2 &  BitsI1)*(-3));

        % Quadrature:
        xQ = ((~BitsQ2 & ~BitsQ1)*(+3) + ( BitsQ2 & ~BitsQ1)*(+1) + ...
               ( BitsQ2 &  BitsQ1)*(-1) + (~BitsQ2 &  BitsQ1)*(-3));

        % Normalized symbols:
        x = (1/sqrt(10))*(xI + xQ*1i);
    end
end
```

Matlab/Octave Code 2.2 Impulse response of RRC filters

```
function g = RRC(Span,SpS,Rolloff)
%%%%%%%%%%%%%%%%%%%%%%%%%%%%%%%%%%%%%%%%%%%%%%%%%%%%%%%%%%%%%%%%%%%%%%%%%%%%
% SRRC g = RRC(Span,SpS,Rolloff)                                         %
%                                                                        %
%  This function generates the impulse response of a root-raised cosine  %
% (RRC) filter with roll-off 'Rolloff', span of 'Span' symbols, and with %
% 'SpS' samples per symbol.                                              %
%                                                                        %
% Input:                                                                 %
%   Span    = Span (in symbols) of the filter;                           %
%   SpS     = Number of samples per symbol to be considered;             %
%   Rolloff = Roll-off of the RRC filter;                                %
%                                                                        %
% Output:                                                                %
%   g = Impulse response of the RRC filter (column vector);              %
%                                                                        %
% This function is part of the book Digital Coherent Optical Systems;    %
% Darli A. A. Mello and Fabio A. Barbosa;                                %
%%%%%%%%%%%%%%%%%%%%%%%%%%%%%%%%%%%%%%%%%%%%%%%%%%%%%%%%%%%%%%%%%%%%%%%%%%%%

    % Configurations:
    g = zeros(Span*SpS+1,1); k = (-Span*SpS/2:Span*SpS/2)/SpS;
    i1 = find(k==0) ; i2 = find(abs(4*Rolloff*k)-1==0) ; i3 = 1:length(k);
    i3([i1 i2]) = []        ; k = k(i3);

    % Singularity in k = 0:
    if ~isempty(i1)
        g(i1) = 1-Rolloff + 4*Rolloff/pi;
    end

    % Singularity in k = 1/(4*Rolloff):
    if ~isempty(i2)
```

```
        g(i2) = Rolloff/sqrt(2)*((1+2/pi)*sin(pi/(4*Rolloff))+...
            (1-2/pi)*cos(pi/(4*Rolloff)));
    end

    % Calculating the coefficients for k ~= 0 and k ~= 1/(4*Rolloff):
    g(i3) = (sin(pi*k.*(1-Rolloff)) + 4*Rolloff*k.*cos(pi*k.*(1+Rolloff)))...
        ./(pi*k.*(1-(4*Rolloff*k).^2));

    % Normalizing the amplitude of the filter:
    g = g/max(g);
end
```

Matlab/Octave Code 2.3 Pulse shaping using an RRC filter

```
function [xb] = PulseShaping(x,SpS,ParamFilter)
%%%%%%%%%%%%%%%%%%%%%%%%%%%%%%%%%%%%%%%%%%%%%%%%%%%%%%%%%%%%%%%%%%%%%%%%%%%%%
% PULSESHAPING xb = PulseShaping(x,SpS,ParamFilter)                        %
%                                                                          %
%   This function performs pulse shaping of a sequence of symbols 'x' using%
% a root-raised cosine (RRC) filter. The filter parameters are defined in  %
% 'ParamFilter'. The sequence of symbols is first upsampled to 'SpS'       %
% samples per symbol and, then, applied to the RRC filter.                 %
%                                                                          %
% Input:                                                                   %
%   x          = Sequence of symbols to be transmitted (one pol.           %
%                orientation - column vector);                             %
%   SpS        = Number of samples per symbol to be considered during      %
%                pulse shaping;                                            %
%   ParamFilter = Struct that specifies parameters of the RRC filter:      %
%            - ParamFilter.Rolloff: Roll-off factor (between 0 and 1);     %
%            - ParamFilter.Span: Filter span (in symbols);                 %
%                                                                          %
% Output:                                                                  %
%   xb = Signal after pulse shaping, normalized to unitary power (column   %
%        vector);                                                          %
%                                                                          %
% This function is part of the book Digital Coherent Optical Systems;      %
% Darli A. A. Mello and Fabio A. Barbosa;                                  %
%%%%%%%%%%%%%%%%%%%%%%%%%%%%%%%%%%%%%%%%%%%%%%%%%%%%%%%%%%%%%%%%%%%%%%%%%%%%%

    % Obtaining the filter transfer function:
    g = RRC(ParamFilter.Span,SpS,ParamFilter.Rolloff);

    % Upsampling the symbols for pulse shaping:
    xUpsamp = upsample(x(:,1),SpS);

    % Filtering the upsampled symbols:
    xb(:,1) = conv(xUpsamp,g,'same');

    % Normalizing the signal to unitary power:
    xb(:,1) = xb(:,1)/sqrt(mean(abs(xb(:,1)).^2));
end
```

Matlab/Octave Code 2.4 Decision of QPSK and 16-QAM symbols considering sequential decoding

```
function [Decided] = Decision(r,ModFormat,BitsOutput)
%%%%%%%%%%%%%%%%%%%%%%%%%%%%%%%%%%%%%%%%%%%%%%%%%%%%%%%%%%%%%%%%%%%%%%%%%%%%%
% DECISION [Decided] = Decision(r,ModFormat,BitsOutput)                    %
%                                                                          %
%   This function performs decisions on each symbol of sequence 'r'        %
% according to the modulation format 'ModFormat'. The function generates   %
% as output the sequence of decided symbols or the sequence of bits        %
% corresponding to the decided symbols (assuming the same binary labeling  %
```

```
% of function 'SymbolGeneration'), depending on the flag 'BitsOutput'.    %
% This function does not perform decisions in the case of differential     %
% encoding.                                                                %
%                                                                          %
% Input:                                                                   %
%    r            = Sequence of symbols received (one pol. orientation -   %
%                   column vector) normalized to unitary power;            %
%    ModFormat    = Modulation format: 'QPSK' or '16-QAM';                 %
%    BitsOutput   = If 'BitsOutput = false', the function generates as output%
%                   the sequence of decided symbols. If 'BitsOutput = true',%
%                   the function generates as output the sequence of bits  %
%                   corresponding to the decided symbols;                  %
%                                                                          %
% Output:                                                                  %
%    Decided = Sequence of decided symbols (if 'BitsOutput = false') or    %
%              the sequence of bits corresponding to the decided symbols   %
%              (if 'BitsOutput = true'). In both cases, 'Decided' is a     %
%              column vector.                                              %
%                                                                          %
% This function is part of the book Digital Coherent Optical Systems;      %
% Darli A. A. Mello and Fabio A. Barbosa;                                  %
%%%%%%%%%%%%%%%%%%%%%%%%%%%%%%%%%%%%%%%%%%%%%%%%%%%%%%%%%%%%%%%%%%%%%%%%%%%%%%

    % Decision:
    switch ModFormat
        case 'QPSK'
            % Decision regions for the in-phase components:
            R1 = real(r) >= 0; R2 = real(r) < 0;
            % Applying the decision regions to the imaginary axis:
            R3 = imag(r) >= 0; R4 = imag(r) < 0;
        case '16QAM'
            % Applying the decision regions to the real axis:
            R1 = real(r) >= 2/sqrt(10); R2 = real(r) >= 0;
            R3 = real(r)  < 0          ; R4 = real(r) <=-2/sqrt(10);
            % Applying the decision regions to the imaginary axis:
            R5 = imag(r) >= 2/sqrt(10); R6 = imag(r) >= 0;
            R7 = imag(r) < 0          ; R8 = imag(r) <=-2/sqrt(10);
        otherwise
            error('The Supported modulation formats are QPSK and 16-QAM;');
    end

    if BitsOutput
        % Binary labeling:
        switch ModFormat
            case 'QPSK'
                ModBits = 2;
                Decided = NaN(length(r),2);
                % Assigning the bits based on the mapping done in the
                % transmitter:
                Decided(R1,2) = zeros(1,sum(R1));
                Decided(R2,2) = ones(1,sum(R2));
                Decided(R3,1) = zeros(1,sum(R3));
                Decided(R4,1) = ones(1,sum(R4));
            case '16QAM'
                ModBits = 4;
                Decided = NaN(length(r),4);
                % Assigning the bits based on the mapping done in the
                % transmitter:
                Decided(R1,[2 4])      = repmat([0 0],sum(R1),1);
                Decided(R2&~R1,[2 4]) = repmat([1 0],sum(R2&~R1),1);
                Decided(R3&~R4,[2 4]) = repmat([1 1],sum(R3&~R4),1);
                Decided(R4,[2 4])      = repmat([0 1],sum(R4),1);
                Decided(R5,[1 3])      = repmat([0 0],sum(R5),1);
                Decided(R6&~R5,[1 3]) = repmat([1 0],sum(R6&~R5),1);
                Decided(R7&~R8,[1 3]) = repmat([1 1],sum(R7&~R8),1);
                Decided(R8,[1 3])      = repmat([0 1],sum(R8),1);
        end
```

```
        % Obtaining the decided bits as a column vector:
        Decided = reshape(Decided',1,length(r)*ModBits)';
    else
        % Decided Symbols:
        Decided = zeros(size(r));
        switch ModFormat
            case 'QPSK'
                % Assigning the bits based on the mapping done in the
                % transmitter:
                Decided(R1) = Decided(R1) + (1/sqrt(2));
                Decided(R2) = Decided(R2) - (1/sqrt(2));
                Decided(R3) = Decided(R3) + (1i*1/sqrt(2));
                Decided(R4) = Decided(R4) - (1i*1/sqrt(2));
            case '16QAM'
                % Assigning the bits based on the mapping done in the
                % transmitter:
                Decided(R1)       = Decided(R1)    + (3/sqrt(10));
                Decided(R2 & ~R1) = Decided(R2&~R1) + (1/sqrt(10));
                Decided(R3 & ~R4) = Decided(R3&~R4) - (1/sqrt(10));
                Decided(R4)       = Decided(R4)    - (3/sqrt(10));
                Decided(R5)       = Decided(R5)    + (3i/sqrt(10));
                Decided(R6 & ~R5) = Decided(R6&~R5) + (1i/sqrt(10));
                Decided(R7 & ~R8) = Decided(R7&~R8) - (1i/sqrt(10));
                Decided(R8)       = Decided(R8)    - (3i/sqrt(10));
        end
    end
end
```

Matlab/Octave Code 2.5 AWGN generation considering a given SNR$_b$ in dB

```
function [r] = NoiseInsertion(x,ModBits,SNRb_dB,SpS)
%%%%%%%%%%%%%%%%%%%%%%%%%%%%%%%%%%%%%%%%%%%%%%%%%%%%%%%%%%%%%%%%%%%%%%%%%
% NOISEINSERTION [r] = NoiseInsertion(x,ModBits,SNRb_dB,SpS)          %
%                                                                     %
%  This function inserts additive white Gaussian noise (AWGN) in the  %
% transmitted signal 'x' (single-polarization orientation), so that an %
% SNR per bit (in dB) 'SNRb_dB' is achieved.                          %
%                                                                     %
% Input:                                                              %
%   x       = Transmitted signal (single pol. orientation - column vector)%
%   ModBits = Number of bits per symbol of the modulation format used in %
%             in the signal 'x';                                      %
%   SNRb_dB = SNR per bit in dB                                       %
%   SpS     = Number of samples per symbol in the input signal 'x';   %
%                                                                     %
% Output:                                                             %
%   r = Signal after noise insertion (column vector);                %
%                                                                     %
% This function is part of the book Digital Coherent Optical Systems; %
% Darli A. A. Mello and Fabio A. Barbosa;                            %
%%%%%%%%%%%%%%%%%%%%%%%%%%%%%%%%%%%%%%%%%%%%%%%%%%%%%%%%%%%%%%%%%%%%%%%%%

    % SNR per bit in linear scale:
    SNRb_Lin = 10^(SNRb_dB/10);

    % AWGN standard deviation:
    StdDev = sqrt(mean(abs(x).^2)*SpS/(2*ModBits*SNRb_Lin));

    % AWGN generation:
    n = StdDev*randn(length(x),1) + 1i*StdDev*randn(length(x),1);

    % Inserting noise to the signal:
    r = x + n;
end
```

2.5.2   *Functions for Sect. 2.2*

Matlab/Octave Code 2.6   Simulates a laser as a continuous wave optical source

```matlab
function [E] = Laser(ParamLaser,SpS,Rs,NSymb,NPol)
%%%%%%%%%%%%%%%%%%%%%%%%%%%%%%%%%%%%%%%%%%%%%%%%%%%%%%%%%%%%%%%%%%%%%%%%%%%%
% LASER [E] = Laser(ParamLaser,SpS,Rs,NSymb,NPol)                        %
%                                                                        %
%  This function simulates a laser as a continuous wave optical source,  %
% according to the parameters defined in 'ParamLaser'. The electric field %
% 'E' of the optical signal has total power defined by 'ParamLaser.Pcw', %
% for both the scenarios with one or two polarization orientations. Phase %
% noise is inserted into the electric field 'E' if the laser linewidth   %
% defined by 'ParamLaser.Linewidth' is different from 0 Hz.              %
%                                                                        %
% Input:                                                                 %
%   SpS       = Number of samples per symbol in the oversampled signal   %
%                 (e.g., at the signal to be transmitted);               %
%   Rs        = Symbol rate in symbols/second;                           %
%   NSymb     = Number of transmitted symbols;                           %
%   NPol      = Number of polarization orientations used;                %
%   ParamLaser = Struct that specifies parameters of the laser:          %
%      -ParamLaser.Pcw: Total power in dBm;                              %
%      -ParamLaser.Linewidth: Laser linewidth in Hz. The default value is %
%              0 Hz (no phase noise). If the laser linewidth is not 0 Hz, %
%              phase noise is inserted into the optical carrier*;        %
%      *Note: 'ParamLaser.Linewidth' should be defined only if the required%
%             value differs from 0 Hz;                                   %
%                                                                        %
% Output:                                                                %
%   E = Electric field of the optical signal. 'E' is a column vector if  %
%       NPol = 1, or a matrix with two column-oriented vectors if NPol = 2, %
%       where each column has the signal of a pol. orientation (V and H pol.%
%       orientations);                                                   %
%                                                                        %
% This function is part of the book Digital Coherent Optical Systems;    %
% Darli A. A. Mello and Fabio A. Barbosa;                                %
%%%%%%%%%%%%%%%%%%%%%%%%%%%%%%%%%%%%%%%%%%%%%%%%%%%%%%%%%%%%%%%%%%%%%%%%%%%%

        % Power of the continuous wave (Pcw) in dBm:
        Pcw = ParamLaser.Pcw;

        % Calculating the linear power of the continuous wave:
        PcwLinear = 1e-3*10^(Pcw/10);

        % Generating the electric filed of the optical signal:
        if NPol == 1
            E = ones(SpS*NSymb,1)*sqrt(PcwLinear);
        elseif NPol == 2
            E = ones(SpS*NSymb,2)*sqrt(PcwLinear/2);
        else
            error('The possible number of polarizations used must be 1 or 2');
        end

        % Laser Linewidth. Note: If the laser linewidth is not 0 Hz, phase
        % noise is inserted into the optical carrier:
        if isfield(ParamLaser,'Linewidth')
            if ParamLaser.Linewidth ~= 0
                % Period between samples at the (oversampled) transmit. signal:
                T = 1/(SpS*Rs);

                % Calculating the phase noise:
                Var         = 2*pi*ParamLaser.Linewidth*T ;
                Delta_theta = sqrt(Var)*randn(size(E,1),1);
                Theta       = cumsum(Delta_theta);
```

```
             % Adding phase noise to the optical signal:
             E = E.*repmat(exp(1i*Theta),1,size(E,2));
       end
    end
end
```

2.5.3   Functions for Sect. 2.3

Matlab/Octave Code 2.7 IQM simulation

```
function [EOutput] = IQModulator(xb,EInput,ParamMZM)
%%%%%%%%%%%%%%%%%%%%%%%%%%%%%%%%%%%%%%%%%%%%%%%%%%%%%%%%%%%%%%%%%%%%%%%%%%%%
% IQMODULATOR [EOutput] = IQModulator(xb,EInput,ParamMZM)                %
%                                                                        %
%  This function simulates an in-phase and quadrature modulator (IQM). The%
% electrical field 'EInput' of the optical signal is modulated according %
% to the signal 'xb', generating 'EOutput'. The Mach--Zehnder modulators %
% (MZMs) that compose the IQM are considered to be identical and have the %
% parameters specified in 'ParamMZM'.                                    %
%                                                                        %
% Input:                                                                 %
%  xb       = (Electric) Signal to be transmitted (in one pol. orientation%
%             - column vector);                                          %
%  EInput   = Optical carrier (single pol. orientation - column vector); %
%  ParamMZM = Struct that specifies parameters of the MZMs that compose  %
%             the IQM:                                                   %
%     -ParamMZM.Vpi    = MZM Vpi;                                        %
%     -ParamMZM.Bias   = Bias voltage;                                   %
%     -ParamMZM.MaxExc = Upper limit for the excursion of the modulation %
%                        signal*;                                        %
%     -ParamMZM.MinExc = Lower limit for the excursion of the modulation %
%                        signal*;                                        %
%       *Note: In this function, the modulation signal is scaled so that %
%              it fits the excursion defined in 'ParamMZM.MaxExc/MinExc'; %
%                                                                        %
% Output:                                                                %
%   EOuput = IQM output signal (column vector);                          %
%                                                                        %
% This function is part of the book Digital Coherent Optical Systems;    %
% Darli A. A. Mello and Fabio A. Barbosa;                                %
%%%%%%%%%%%%%%%%%%%%%%%%%%%%%%%%%%%%%%%%%%%%%%%%%%%%%%%%%%%%%%%%%%%%%%%%%%%%

    % Obtaining the in-phase and quadrature components of the electrical:
    mI = real(xb); mI = mI/max(abs(mI)); % In-phase;
    mQ = imag(xb); mQ = mQ/max(abs(mQ)); % Quadrature;

    % Setting the signal excursion:
    mI = mI*(ParamMZM.MaxExc-ParamMZM.MinExc)/2;
    mQ = mQ*(ParamMZM.MaxExc-ParamMZM.MinExc)/2;

    % Obtaining the signals after considering the bias:
    vI = mI+ParamMZM.Bias; vQ = mQ+ParamMZM.Bias;

    % Phase modulation in the in-phase and quadrature branches;
    PhiI = pi*(vI)/ParamMZM.Vpi; PhiQ = pi*(vQ)/ParamMZM.Vpi;

    % IQM output signal:
    EOutput = (0.5*cos(0.5*PhiI) + 0.5i*cos(0.5*PhiQ)).*EInput;
end
```

References

1. P.J. Winzer, R.J. Essiambre, Advanced optical modulation formats. Proc. IEEE **94**(5), 952–985 (2006)
2. R.J. Essiambre, G. Kramer, P.J. Winzer, G.J. Foschini, B. Goebel, Capacity limits of optical fiber networks. J. Lightwave Technol. **28**(4), 662–701 (2010)
3. J.R. Barry, D.G. Messerschmitt, E.A. Lee, *Digital Communication*, 3rd edn. (Kluwer Academic Publishers, Norwell, MA, USA, 2004)
4. A. Shahpari, R.M. Ferreira, R.S. Luis, Z. Vujicic, F.P. Guiomar, J.D. Reis, A.L. Teixeira, Coherent access: a review. J. Lightwave Technol. **35**(4), 1050–1058 (2017)
5. K.-P. Ho, *Phase-Modulated Optical Communication Systems* (Springer, Boston, MA, 2005)
6. H. Nyquist, Certain topics in telegraph transmission theory. Trans. Am. Inst. Electr. Eng. **47**(2), 617–644 (1928)
7. J.R. Barry, E.A. Lee, Performance of coherent optical receivers. Proc. IEEE **78**(8), 1369–1394 (1990)
8. S. Haykin, *Communication Systems*, 4th edn. (Wiley, New York, USA, 2001)
9. A.L. Schawlow, C.H. Townes, Infrared and optical masers. Phys. Rev. **112**, 1940–1949 (1958)
10. T. Mainman, Stimulated optical radiation in ruby. Nature **187**, 493–494 (1960)
11. A. Einstein, Zur quantentheorie der strahlung. Physikalische Zeitschrift **18**, 121–128 (1917)
12. G.P. Agrawal, *Fiber-Optic Communication Systems*, 4th edn. (Wiley, Hoboken, NJ, 2010)
13. M. Pollnau, Phase aspect in photon emission and absorption. Optica **5**(4), 465–474 (2018)
14. H. Kogelnik, C.V. Shank, Coupled-wave theory of distributed feedback lasers. J. Appl. Phys. **43**(5), 2327–2335 (1972)
15. R. Ramaswami, K. Sivarajan, G. Sasaki, *Optical Networks: A Practical Perspective*, 3rd edn. (Morgan Kaufmann Publishers, San Francisco, CA, USA, 2009)
16. L. Thylén, U. Westergren, P. Holmström, R. Schatz, P. Jänes, Recent developments in high-speed optical modulators, in *Optical Fiber Telecommunications V A*, ed. by I.P. Kaminow, T. Li, A.E. Willner. Optics and Photonics, 5th edn. (Academic Press, Burlington, 2008), pp. 183–220. [Online]. Available: http://www.sciencedirect.com/science/article/pii/B9780123741714000071
17. M. Seimetz, *High-Order Modulation for Optical Fiber Transmission*, ser. Springer Series in Optical Sciences, vol. 143. (Springer, Berlin, Heidelberg, 2009)

Chapter 3
The Optical Channel

The optical communication channel is the result of several interactions between optical signals and matter. These effects can occur in the various fiber-optic sections of the link or in the devices traversed by the optical signal. We will not address all of these effects, but only those most significant for the modeling of digital coherent optical systems. The most fundamental effect for any communication system is attenuation, which reduces the signal power along the link as a result of absorption and scattering phenomena. Because of attenuation, the transmitted signals need to be periodically amplified by in-line amplifiers, which insert noise and limit the system reach. Polarization mode dispersion (PMD) and chromatic dispersion (CD) are two important linear effects that spread symbols over time and cause intersymbol interference. They are usually compensated for at the receiver using digital signal processing (DSP) techniques. Although PMD and CD individually do not limit capacity, they influence the complexity of the receiver DSP, as CD may require very long static equalizers and PMD demands adaptive equalization.

In the class of nonlinear fiber effects, those originating from the Kerr effect, particularly self-phase modulation (SPM) and cross-phase modulation (XPM), are the most prominent. SPM generates distortions in a certain channel due to nonlinear interactions between electromagnetic waves in this same channel with the fiber material. In XPM, however, the nonlinear interference is generated by the interaction of electromagnetic waves in two different channels and the fiber material. In principle, it is possible to compensate for SPM using DSP techniques such as digital backpropagation (DBP), which are based on the computational reverse propagation of the received signal, in an attempt to recover the transmitted one [1, 2]. The performance of DBP is limited in several ways, including the accumulation of noise, uncertainty in the link parameters, and, mainly, the dominance of XPM. Although DBP generates tangible gains in single-channel systems, where SPM is the dominant effect, in a fully populated spectrum XPM is dominant, and DBP provides reduced gains [3, 4]. A DBP algorithm for XPM is in principle also possible [5], but it would require the processing of multiple wavelength channels by the same digital

© Springer Nature Switzerland AG 2021
D. A. de Arruda Mello, F. A. Barbosa, *Digital Coherent Optical Systems*,
Optical Networks, https://doi.org/10.1007/978-3-030-66541-8_3

signal processor, which is not the current standard practice in wavelength division multiplexing (WDM) transmission systems. The impact of the Kerr effect and its consequences as XPM and SPM on the system performance depend intrinsically on the CD management along the link. CD causes pulse broadening in tens or hundreds of adjacent symbols, modifying the temporal distribution of the transmitted signal power and the nonlinear relationship of different WDM channels. Therefore, modern coherent optical systems with fully digital CD compensation exhibit a very different nonlinear behavior compared with former optical systems with in-line optical CD compensation. In the absence of inline optical CD compensation modules in these systems, the effect of nonlinear interference in a long-range multichannel system becomes virtually equivalent to that of additive white Gaussian noise (AWGN) [6–8]. In this case, the nonlinear interference adds up with amplifier noise as just another AWGN contribution. The next sections discuss these effects in more detail.

3.1   Propagation Modes and Single-Mode Condition

As we have seen in Chap. 1, since the emergence of the second generation of fiber-optic communication systems, long-distance transmission has relied on single-mode fibers (SMFs). It is true that the sixth generation of optical systems based on space-division multiplexing (SDM) may rely on multimode propagation, but this is still uncertain. Thus, to study the main effects in long-distance transmission, we have to first understand single-mode propagation.

An optical fiber is basically a cylindrical dielectric waveguide made up essentially of two layers, called the core and the cladding. The effective guiding of electromagnetic radiation requires the refractive index in the core to be higher than in the cladding. In order to understand the modes supported by an optical fiber, let us recall the four Maxwell's equations for the propagation of electromagnetic waves in a dielectric medium, i.e., free of charges

$$\nabla \cdot \mathbf{D} = 0 \qquad \text{(Gauss's Law)}, \tag{3.1}$$

$$\nabla \cdot \mathbf{B} = 0 \qquad \text{(Gauss's Law for Magnetism)}, \tag{3.2}$$

$$\nabla \times \mathbf{H} = \frac{\partial \mathbf{D}}{\partial t} \qquad \text{(Ampere–Maxwell's Law)}, \tag{3.3}$$

$$\nabla \times \mathbf{E} = -\frac{\partial \mathbf{B}}{\partial t} \qquad \text{(Faraday's Law)}, \tag{3.4}$$

where $\mathbf{E}$ (V/m) is the electric field intensity and $\mathbf{H}$ (A/m) is the magnetic field intensity. The electric flux density $\mathbf{D}$ (C/m^2) and the magnetic flux density $\mathbf{B}$ (T) are related to $\mathbf{E}$ and $\mathbf{H}$ through the constitutive relations

$$\mathbf{D} = \epsilon_0 \mathbf{E} + \mathbf{P}, \tag{3.5}$$

$$\mathbf{B} = \mu_0 \mathbf{H} + \mathbf{M}. \tag{3.6}$$

Constants ϵ_0 and μ_0 are the vacuum electric permittivity and the vacuum magnetic permeability, respectively, given by

$$\epsilon_0 = 8.854 \times 10^{-12} \text{ F/m}, \tag{3.7}$$

$$\mu_0 = 4\pi \times 10^{-7} \text{ H/m}. \tag{3.8}$$

Fields $\mathbf{P}$ and $\mathbf{M}$ are the electric and magnetic polarization generated by the medium as a response to the applied fields.

At this point, it is convenient to solve Maxwell's equations in frequency domain, which simplifies the time derivatives. Recalling the definition of the Fourier transform

$$\mathfrak{F}\{\mathbf{E}\} = \tilde{\mathbf{E}} = \int_{-\infty}^{+\infty} \mathbf{E} e^{-j\omega t} \, dt, \tag{3.9}$$

the inverse Fourier transform is given by

$$\mathbf{E} = \frac{1}{2\pi} \int_{-\infty}^{+\infty} \tilde{\mathbf{E}} e^{j\omega t} \, d\omega. \tag{3.10}$$

Differentiating both sides with respect to t yields

$$\frac{\partial \mathbf{E}}{\partial t} = \frac{1}{2\pi} \int_{-\infty}^{+\infty} j\omega \tilde{\mathbf{E}} e^{j\omega t} \, d\omega, \tag{3.11}$$

i.e., the inverse transform of $j\omega\tilde{\mathbf{E}}$ is equal to the temporal derivative of $\mathbf{E}$. Then, applying the Fourier transform to both sides of (3.11)

$$\mathfrak{F}\left\{\frac{\partial \mathbf{E}}{\partial t}\right\} = j\omega\tilde{\mathbf{E}}. \tag{3.12}$$

Thus, in frequency domain, the four Maxwell's equations become

$$\nabla \cdot \tilde{\mathbf{D}} = 0, \tag{3.13}$$

$$\nabla \cdot \tilde{\mathbf{B}} = 0, \tag{3.14}$$

$$\nabla \times \tilde{\mathbf{H}} = j\omega\tilde{\mathbf{D}}, \tag{3.15}$$

$$\nabla \times \tilde{\mathbf{E}} = -j\omega\tilde{\mathbf{B}}. \tag{3.16}$$

Likewise, in frequency domain, the constitutive relations in a homogeneous medium become

$$\tilde{\mathbf{D}} = \epsilon_0 \tilde{\mathbf{E}} + \tilde{\mathbf{P}}, \tag{3.17}$$

$$\tilde{\mathbf{B}} = \mu_0 \tilde{\mathbf{H}} + \tilde{\mathbf{M}}. \tag{3.18}$$

In optical fibers made of silica glasses, $\tilde{\mathbf{M}} = 0$, and the electric polarization $\tilde{\mathbf{P}}$ is proportional to the applied electric field intensity $\tilde{\mathbf{E}}$. Assuming a homogeneous medium, the proportionality constant is given by $\epsilon_0 \chi^{(1)}$, i.e., $\mathbf{P} = \epsilon_0 \chi^{(1)} \mathbf{E}$, where $\chi^{(1)}$ is the linear susceptibility of the medium. Although $\chi^{(1)}$ is frequency dependent, for the sake of clarity, we omit this dependence for the moment. Thus, the constitutive relations become

$$\tilde{\mathbf{D}} = \epsilon_0 \tilde{\mathbf{E}} + \tilde{\mathbf{P}} = \epsilon_0 (1 + \chi^{(1)}) \tilde{\mathbf{E}} = \epsilon \tilde{\mathbf{E}}, \tag{3.19}$$

$$\tilde{\mathbf{B}} = \mu_0 \tilde{\mathbf{H}} = \mu \tilde{\mathbf{H}}, \tag{3.20}$$

where, in this case, $\epsilon = (1 + \chi^{(1)})$ and $\mu = \mu_0$. In Maxwell's equations, the quantities related to the electric field $\mathbf{E}$ and the magnetic field $\mathbf{H}$ are coupled. They are usually decoupled by first applying the rotational to both sides of Faraday's law

$$\nabla \times \nabla \times \tilde{\mathbf{E}} = -j\omega \nabla \times \tilde{\mathbf{B}}. \tag{3.21}$$

Then, the identity $\nabla \times \nabla \times \mathbf{A} = \nabla(\nabla \cdot \mathbf{A}) - \nabla^2 \mathbf{A}$ is applied. In a dielectric medium, $\nabla \cdot \tilde{\mathbf{E}} = 0$, and therefore

$$\nabla^2 \tilde{\mathbf{E}} + \omega^2 \mu \epsilon \tilde{\mathbf{E}} = 0. \tag{3.22}$$

This is the important wave equation that governs the propagation of electromagnetic waves. The same expression can be derived for $\tilde{\mathbf{H}}$. Solving for the full electromagnetic field means finding components $\tilde{E}_x$, $\tilde{E}_y$, $\tilde{E}_z$, $\tilde{H}_x$, $\tilde{H}_y$, and $\tilde{H}_z$. Since the optical fiber has a cylindrical profile, we can use cylindrical coordinates, so that each component will depend on the radial component r, the azimuthal component ϕ, and the longitudinal component z, as indicated in Fig. 3.1 [9]. For a given frequency component ω, the following expressions are obtained for $\tilde{\mathbf{E}}$ and $\tilde{\mathbf{H}}$:

$$\tilde{\mathbf{E}}(r, \phi, z) = \tilde{E}_x(r, \phi, z)\mathbf{x} + \tilde{E}_y(r, \phi, z)\mathbf{y} + \tilde{E}_z(r, \phi, z)\mathbf{z}, \tag{3.23}$$

$$\tilde{\mathbf{H}}(r, \phi, z) = \tilde{H}_x(r, \phi, z)\mathbf{x} + \tilde{H}_y(r, \phi, z)\mathbf{y} + \tilde{H}_z(r, \phi, z)\mathbf{z}, \tag{3.24}$$

where $\mathbf{x}$, $\mathbf{y}$, and $\mathbf{z}$ are unit vectors pointing toward directions x, y, and z. Each of the six components can be further expressed as the product of independent functions of r, ϕ, and z, i.e., $F(r, \phi, z) = R(r) \times \Phi(\phi) \times Z(z)$. As the solutions of the wave equations are given in terms of propagating fields, we will assume the dependence of $\tilde{\mathbf{E}}$ and $\tilde{\mathbf{H}}$ with respect to z as $Z(z) = e^{-j\beta(\omega)z}$, where $\beta = 2\pi/\lambda$ is the modal propagation constant. In the solution of guided modes in long waveguides, considered infinite for the mode calculations, it is usual to first solve for $\tilde{E}_z$ and $\tilde{H}_z$ and then obtain the remaining four field components by applying the boundary

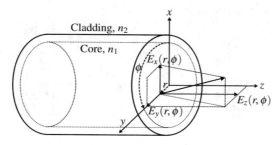

Fig. 3.1 Fiber optic as a cylindrical waveguide. A cylindrical core with refractive index n_1 is surrounded by a cladding with refractive index n_2. The figure also indicates the system of coordinates used for mode calculations, where the coordinates of a specific point in space are described by its cylindrical coordinates r, ϕ, and z, and the vector field consists of components E_x, E_y, and E_z

conditions at the core–cladding interface. In this chapter, we consider a step-index profile for the optical fiber. We assume that the fiber is a cylindrical structure constructed by a central structure called core with refractive index n_1, surrounded by an external structure called cladding with refractive index n_2, as shown in Fig. 3.1.

A very useful and practical simplification makes use of the fact that the refractive indexes of a step-index optical fiber are very close, leading to weakly guided modes. This approximation enables the so-called linearly polarized modes, or LP modes, for which $\tilde{E}_z$ and $\tilde{H}_z$ are small compared to the transverse components [10]. In addition, under the LP approximation, one of the components $\tilde{E}_x$ or $\tilde{E}_y$ can be set to zero, generating a linearly polarized signal. Without loss of generality, let us assume $\tilde{E}_x = 0$, such that $\tilde{E}_y(r, \phi)$ and $\tilde{H}_x(r, \phi)$ are approximately given by Gloge [10]

$$\tilde{E}_y(r, \phi) = \tilde{H}_x(r, \phi) \begin{Bmatrix} Z_0/n_1 \\ Z_0/n_2 \end{Bmatrix} = E_l \begin{Bmatrix} J_l(ur/a)/J_l(u) \\ K_l(wr/a)/K_l(w) \end{Bmatrix} \cos(\phi l), \qquad (3.25)$$

where a is the core radius. The upper expression corresponds to the field in the core and the lower expression to the field in the cladding. Solutions are also supported with $\sin(\phi l)$ in addition to $\cos(\phi l)$. J_l is the Bessel function of the first kind, and K_l is the modified Bessel function of the second kind. Parameter $Z_0 \approx 377\Omega$ is the impedance of free space, and l is a non-negative integer that indicates the number of oscillations in the azimuthal direction. E_l is the electric field intensity at the core–cladding interface. Constants u and w are related to the properties of the cladding and the core as

$$u = a(k^2 n_1^2 - \beta^2)^{\frac{1}{2}}, \qquad (3.26)$$

$$w = a(\beta^2 - k^2 n_2^2)^{\frac{1}{2}}, \qquad (3.27)$$

Table 3.1 Roots of the Bessel function of the first kind J_l

m	$J_0(x)$	$J_1(x)$	$J_2(x)$	$J_3(x)$	$J_4(x)$	$J_5(x)$
1	2.4048	3.8317	5.1356	6.3802	7.5883	8.7715
2	5.5201	7.0156	8.4172	9.7610	11.0647	12.3386
3	8.6537	10.1735	11.6198	13.0152	14.3725	15.7002
4	11.7915	13.3237	14.7960	16.2235	17.6160	18.9801
5	14.9309	16.4706	17.9598	19.4094	20.8269	22.2178

where $k = 2\pi/\lambda$ is the wave number (λ is the wavelength in free space), and $\beta = 2\pi/\lambda_g$ is the propagation constant of the guided mode (λ_g is the wavelength for the guided mode).

We have mentioned the propagation constant β, but we did not discuss how to calculate it. It is usually obtained from the boundary conditions of tangential fields at the core–cladding interface. Under the weakly guiding approximation, the boundary condition reduces to [10]

$$u[J_{l-1}(u)/J_l(u)] = -w[(K_{l-1}(w)/K_l(w))]. \tag{3.28}$$

Another useful parameter is the so-called normalized frequency v, defined as

$$v = \sqrt{u^2 + w^2} = ak\sqrt{n_1^2 - n_2^2}. \tag{3.29}$$

For a given normalized frequency v, there is a finite number of solutions of (3.28), and each one corresponds to a different propagation constant β. These solutions are the supported *propagation modes*. These linearly polarized modes are called LP_{lm}, l being the number of oscillations in the azimuthal direction ϕ, and m the index of the solution. For example, $LP_{1,2}$ is the second solution of a mode with one cycle of sinusoidal variation in the azimuthal direction ϕ. The minimum frequencies supported by each mode (called cut-off frequencies) are given by making $w = 0$ in (3.28). As K_l is positive, $v = u$ and $J_{l-1} = 0$. Figure 3.2 shows $J_{l-1}(v)$ for $l = 1$ and $l = 0$. For $l = 0$, we use the relationship $J_{-1} = -J_1$. Mode LP_{01}, called fundamental mode, is always supported as, for $v = 0$, $J_{-1} = 0$, and $J_0 \neq 0$. The single-mode condition implies $v < v_c^{11} = 2.405$, such that LP_{01} is the only supported mode. Table 3.1 contains the roots of the Bessel function of the first kind J_l. The first root corresponds to mode LP_{11} (2.4048), followed by LP_{21} and LP_{02} (3.8317), LP_{31} (5.1356), LP_{12} (5.5201), and so on.

Figure 3.3 shows the modal fields for several LP_{lm} modes calculated according to (3.25). The LP_{01} (fundamental mode) and LP_{02} modes do not oscillate in the azimuthal direction, as $l = 0$. The fundamental mode has a monotonic decreasing field in the radial direction, while the second solution LP_{02} has one minimum. The LP_{lm} modes with $l > 0$ have two degenerate solutions, corresponding to the $\cos(\phi l)$ and $\sin(\phi l)$ solutions in (3.25). Higher-order modes with $l = 2, 3$ are also shown in the figure.

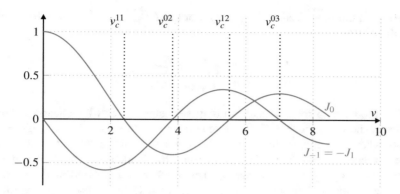

Fig. 3.2 Cut-off frequencies for LP_{lm} modes, for $l = 0$ and $l = 1$. The fundamental mode LP_{01} does not have a cut-off frequency. Single-mode condition requires $v < v_c^{11} = 2.405$

Fig. 3.3 Modal fields of LP_{lm} modes in the x or y polarization, calculated according to (3.25). The fields plotted side-by-side correspond to the $\cos(\phi l)$ and $\sin(\phi l)$ solutions

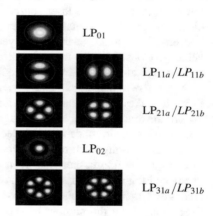

LP$_{01}$

LP$_{11a}$/LP$_{11b}$

LP$_{21a}$/LP$_{21b}$

LP$_{02}$

LP$_{31a}$/LP$_{31b}$

A useful tool for visualizing the generation of modes with increasing frequencies is the so-called modal map, which plots normalized propagation constants b, defined as

$$b(v) = 1 - (u^2/v^2) \approx \frac{\beta/k - n_2}{n_1 - n_2}, \tag{3.30}$$

where the approximation is valid for small index differences. Resorting again to the weakly guiding approximation, one can use a simplified assumption for u of all modes except LP_{01} [10]

$$u(v) = v_c e^{\{[\arcsin(s/v_c) - \arcsin(s/v)]/s\}}, \tag{3.31}$$

$$s = (v_c^2 - l^2 - 1)^{\frac{1}{2}}, \tag{3.32}$$

where v_c is the mode cut-off frequency. For the LP_{01} mode, $u(v)$ can be obtained by Gloge [10]

$$u(v) = \frac{(1 + \sqrt{2})v}{1 + (4 + v^4)^{\frac{1}{4}}}.$$

(3.33)

The modal map for a weakly guiding step-index fiber with linearly polarized modes is shown in Fig. 3.4. Each curve corresponds to a specific LP_{lm} mode. The points where the curves intersect the $y = 0$ line indicate the cut-off frequency for each mode. As we have discussed before, the fundamental mode is supported regardless of the operating frequency. The number of supported modes increases with the normalized frequency v in a piecewise fashion. The vast majority of fibers currently used for intra-office operation are multimode. The main advantage of multimode fibers is that their large core radius facilitates the coupling of power generated by the laser source, reducing costs. On the other hand, its main disadvantage is that signals launched into the fiber excite multiple modes, and the portion of the signal energy conveyed by each mode travels with its own group velocity. Consequently, multiple attenuated copies of the same signal arrive at the receiver at different instants, leading to pulse spreading. This phenomenon is known as modal dispersion. Although modal dispersion is tolerable in low-cost fiber links, it makes long-distance transmission using current transceivers impractical. To circumvent this problem, already the second generation of optical systems operate over single-mode fibers. Indeed, operating fibers in the single-mode regime completely eliminates modal dispersion and increases the achievable data rates. Looking at the future, there has been a considerable research effort in recent times to enable long-distance transmission with multimode propagation. The idea of these recent efforts is to exploit the different orthogonal modes to actually multiply the fiber capacity by the number of supported modes. In this case, the separation of the transmitted signals at the receiver is accomplished by complex DSP algorithms. This book is focused on long-distance transmission systems. Therefore, from this section on, we will limit our analyses to single-mode fiber systems.

3.2   Chromatic Dispersion

Before examining the effect of CD, it is appropriate to recall the concepts of phase velocity and group velocity. Phase velocity is the speed of wave crests of a propagating wave. Phase velocity does not indicate the speed of energy or information transfer, but only the speed of a wave pattern. For a mode with propagation constant β, the phase velocity v_p of a sinusoidal signal with frequency ω is given by

$$v_p = \frac{\omega}{\beta}.$$

(3.34)

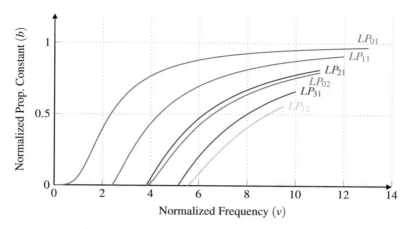

Fig. 3.4 Modal map for a step-index fiber with linearly polarized modes

However, in this book we are interested in conveying information through optical fibers. Information is only conveyed by packets of frequencies, forming a pattern that travels over the fiber with group velocity v_g

$$v_g = \left(\frac{d\beta}{d\omega}\right)^{-1}. \tag{3.35}$$

If the propagation constant has a linear dependency with frequency, $v_g = v_p$, any group of propagating frequencies have the same group velocity, and pulses are not distorted. This is the case, for example, of plane waves traveling in free space. However, in practical waveguides the propagation constant β has a nonlinear dependence on ω, as indicated by the modal map in Fig. 3.4. Thus, different spectral components of a guided mode propagate with distinct group velocities, giving rise to the phenomenon known as CD. If left uncompensated, CD broadens the propagating pulses and leads to intersymbol interference (ISI), as shown in Fig. 3.5. In optical fibers, CD is generated by two main contributions. The first one, called *material dispersion*, is related to the material from which the fiber is made. The second one, called *waveguide dispersion*, arises from the fact that part of the energy of the transmitted signal propagates in the fiber core, and another part in the fiber cladding. As a consequence, the effective refractive index of the mode has an intermediate value, which depends on the energy distribution between the core and the cladding. As this distribution depends on the wavelength, CD is generated.

In a lossless waveguide, the components of an electric or magnetic field in frequency ω vary in the propagation direction z according to

$$E(z, \omega) = E(0, \omega)e^{-j\beta(\omega)z}, \tag{3.36}$$

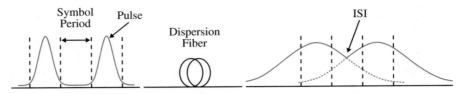

Fig. 3.5 Illustrative example of the effect of CD on the signal propagation. CD spreads pulses in time, leading to intersymbol interference (ISI)

where $\beta(\omega)$ is the propagation constant at ω. The same analysis can be carried out for the frequency components of the spectrum $\tilde{A}(z, \omega)$ of signal $A(z, t)$

$$\tilde{A}(z, \omega) = \tilde{A}(0, \omega)e^{-j\beta(\omega)z}, \qquad (3.37)$$

where $\tilde{A}(z, \omega) = \mathbb{F}\{A(z, t)\}$. Expanding $\beta(\omega)$ in Taylor series around the carrier angular frequency ω_c yields

$$\tilde{A}(z, \omega) = \tilde{A}(0, \omega)e^{-\left(j\beta_0 + j\beta_1 \Delta\omega + \frac{j\beta_2}{2} \Delta\omega^2 + \dots\right)z}, \qquad (3.38)$$

where β_i is the ith derivative of β with respect to ω, and $\Delta\omega = \omega - \omega_c$. Term $e^{j\beta_0 z}$ is a simple phase shift. The multiplication by $e^{j\beta_1 \Delta\omega z}$ generates a time shift of an undistorted pulse. Chromatic dispersion is related to term $e^{\frac{j\beta_2}{2} \Delta\omega^2 z}$. For an observer traveling with the pulse, the spectrum of a signal corrupted by CD can be computed as

$$\tilde{A}(z, \omega) = \tilde{A}(0, \omega)e^{-\frac{j\beta_2 z}{2} \Delta\omega^2}. \qquad (3.39)$$

Parameter β_2 has units of s/m/[rad/s] and gives the spreading of pulses (in seconds) per unit of fiber length (in meters) and per unit of spectral bandwidth (in radians per second). In the specification of optical fibers, however, dispersion is usually quantified by the group velocity dispersion (GVD) parameter, D, in units of s/m², or, better, in ps/nm/km. The GVD parameter indicates the pulse spreading (in picoseconds), per unit of distance (in kilometers), per unit of spectral bandwidth (in nanometers). The parameters D and β_2 are related by

$$D = -\frac{2\pi c}{\lambda^2}\beta_2, \qquad (3.40)$$

where c is the speed of light in vacuum and λ is the operating wavelength. Figure 3.6a illustrates the CD profile for a standard single-mode fiber (SSMF) as a function of the wavelength [11]. The material dispersion has a positive slope, and the waveguide dispersion a negative slope. The combination of both contributions is an increasing profile that achieves $D = 0$ near 1300 nm, and

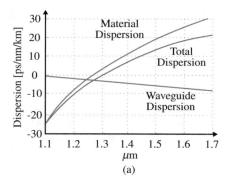 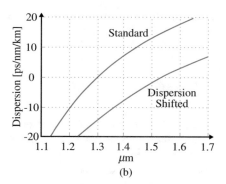

Fig. 3.6 (**a**) Dispersion profile of SSMFs and the contributions of material dispersion and waveguide dispersion. (**b**) Dispersion profile of SSMFs and DSFs

approximately 17 ps/nm/km around $\lambda = 1550$ nm. This is the dispersion figure considered in most operational long-distance links. Figure 3.6b illustrates the dispersion profile of SSMFs and dispersion-shifted (DS) fibers. In the 1980s, DS fibers were developed to shift the zero dispersion wavelength from 1300 to 1550 nm, in order to avoid dispersion in the link. Before the invention of erbium-doped fiber amplifiers (EDFAs) and the deployment of WDM, single-channel systems operated at relatively low power levels. The problem with DS fibers is that their small core area enhances nonlinear effects at high power levels. In WDM systems, the total power is multiplied by the number of channels, enhancing nonlinearities and rendering DS fibers unsuitable for high rates and long distances. As of today, SSMFs and large-effective area fibers are the preferred transmission choices.

The fact that the different spectral components of an optical signal propagate with different group velocities leads to the temporal broadening of pulses. Assuming a linear dependence of group velocity with wavelength, the difference in propagation time between the frequency components in the edges of the signal spectrum is given by

$$\Delta T = |D| L \Delta \lambda, \tag{3.41}$$

where $\Delta \lambda$ is the signal bandwidth.

In long-distance optical transmission, dispersed pulses can spread over hundreds or thousands of symbols. As a rule of thumb, this spread, in number of symbols per kilometer, can be estimated as

$$N \approx \frac{\Delta T}{L T_s} = \frac{|D| \Delta \lambda}{T_s}, \tag{3.42}$$

where T_s is the symbol period. For example, for a minimum-bandwidth Nyquist-shaped signal at a symbol rate of 50 GBd, the spectral width $\Delta \lambda$ is approximately 0.4 nm at $\lambda = 1550$ nm. For SSMF transmission with D = 17 ps/nm/km, N is 0.34

symbols/km. This means that, in a 1000-km fiber system, pulses spread along approximately 340 symbols.

3.3 Polarization Effects

Single-mode fibers support the propagation of two orthogonally polarized modes. Ideally, these modes have the same propagation constant and are, therefore, called degenerate. However, the non-perfect circularity of the core and cladding of optical fibers caused by the manufacturing or installation processes cause the degenerate characteristic of the orthogonally polarized modes to be lost. The fiber then acquires birefringence, which consists of a difference between the refractive indexes (or propagation constants) associated with the polarization modes. In optical fibers, birefringence causes the temporal spread of pulses, a phenomenon called polarization-mode dispersion (PMD). Since pulses launched into the orthogonal polarizations acquire distinct group velocities, it is possible to calculate the difference between the time of arrival of each of them. Linearly birefringent elements have two orthogonal states of polarization with slightly different refractive indexes, called principal states of polarization (PSPs) [12]. Projections of a linearly polarized signal onto the PSPs of the birefringent element experience a mutual delay, called differential group delay (DGD), given by

$$\tau = \frac{l\,\Delta n}{c} = \frac{l\,\Delta\beta}{\omega_0}, \tag{3.43}$$

where Δn and $\Delta\beta$ are the differences between the PSPs' refractive indexes and propagation constants, l is the length of the birefringent element, and ω_0 is the angular carrier frequency.

Polarization effects in optical communications are usually mathematically modeled by Jones calculus, in which the orthogonal polarization components are represented by the elements of a column vector

$$\vec{s} = \begin{bmatrix} E_x \\ E_y \end{bmatrix}, \tag{3.44}$$

where E_x and E_y are the complex field representation in the x and y polarization orientations. The operation of linear optical elements in a certain frequency component ω is represented by a Jones matrix $\mathbf{H}(\omega)$, such that the input–output relationship is given by

$$\begin{bmatrix} E_{o,x}(\omega) \\ E_{o,y}(\omega) \end{bmatrix} = \mathbf{H}(\omega) \begin{bmatrix} E_{i,x}(\omega) \\ E_{i,y}(\omega) \end{bmatrix}, \tag{3.45}$$

where $E_{i,x}(\omega)$, $E_{i,y}(\omega)$, $E_{o,x}(\omega)$, and $E_{o,y}(\omega)$ are the input and output fields in the x and y polarization orientations.

The transfer matrix of a birefringent fiber section i, $\mathbf{H}^{(i)}(\omega)$, can be expressed as [13]

$$\mathbf{H}^{(i)}(\omega) = \mathbf{V}^{(i)}\mathbf{\Lambda}^{(i)}(\omega)\mathbf{U}^{(i)H}, \tag{3.46}$$

where $\mathbf{V}^{(i)}$ and $\mathbf{U}^{(i)}$ are 2×2 complex unitary matrices that describe polarization coupling.[1] In simulations, they can be generated by applying the Gram–Schmidt process to matrices constructed from random complex Gaussian vectors [14]. Alternatively, they can be easily generated by using the $\mathbf{V}^{(i)}$ and $\mathbf{U}^{(i)}$ matrices obtained by the singular value decomposition (SVD) of a $\mathbf{G}^{(i)}$ matrix whose four components are complex Gaussian random variables. The diagonal matrix $\mathbf{\Lambda}^{(i)}(\omega)$ is given by

$$\mathbf{\Lambda}^{(i)}(\omega) = \begin{bmatrix} e^{j\omega\tau^{(i)}/2} & 0 \\ 0 & e^{-j\omega\tau^{(i)}/2} \end{bmatrix}, \tag{3.47}$$

where $\tau^{(i)} = l_i\Delta\beta/\omega_0$ is the DGD caused by birefringence, where l_i is the length of the birefringent element.

Long optical fibers are usually modeled as a concatenation of N birefringent elements, in which the fast and slow polarization axes are rotated with respect to each other, as depicted in Fig. 3.7. Mathematically, this model generates an overall transfer matrix given by the product of the individual elements [13]

$$\mathbf{H}^{(t)}(\omega) = \prod_{i=1}^{N} \mathbf{H}^{(i)}(\omega) = \mathbf{V}^{(t)}(\omega)\mathbf{\Lambda}^{(t)}(\omega)\mathbf{U}^{(t)H}(\omega) = \begin{bmatrix} H_{xx}(\omega) & H_{yx}(\omega) \\ H_{xy}(\omega) & H_{yy}(\omega) \end{bmatrix}, \tag{3.48}$$

where the superscript $(\cdot)^{(t)}$ denotes the total effect obtained by multiplying the individual transfer matrices. Here, $H_{xx}(\omega)$ represents the influence of the input signal in polarization x on the output signal in polarization x, $H_{xy}(\omega)$ the influence of the input signal in polarization x on the output signal in polarization y, $H_{yy}(\omega)$ the influence of the input signal in polarization y on the output signal in polarization y and, finally, $H_{yx}(\omega)$ the influence of the input signal in polarization y on the output signal in the polarization x. The delay matrix $\mathbf{\Lambda}^{(t)}(\omega)$ is given by

$$\mathbf{\Lambda}^{(t)}(\omega) = \begin{bmatrix} e^{j\omega\tau/2} & 0 \\ 0 & e^{-j\omega\tau/2} \end{bmatrix}, \tag{3.49}$$

---

[1]The superscript $(\cdot)^H$ denotes matrix Hermitian conjugate. A complex square matrix $\mathbf{A}$ is unitary if its Hermitian conjugate $\mathbf{A}^H$ is also its inverse, i.e., $\mathbf{A}\mathbf{A}^H = \mathbf{I}$.

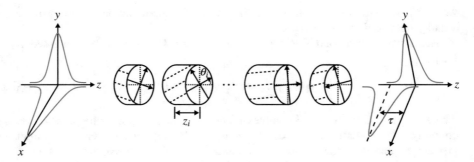

Fig. 3.7 Modeling of PMD by a cascade of sections. Each section i represents a birefringent element of length l_i. In the model, the slow and fast axes in one section are randomly rotated with respect to the other sections. Although the differential group delay (DGD) in each section can be assumed to be constant over time, random coupling produces an accumulated delay τ in the end of the fiber that is a random variable

where τ is the overall DGD at the end of the fiber. It can be calculated from the transfer matrix $\mathbf{H}^{(t)}(\omega)$ through the eigenvalues of a group delay operator defined as [13, 15]

$$j \frac{\partial \mathbf{H}^{(t)}(\omega)}{\partial \omega} \mathbf{H}^{(t)-1}(\omega). \tag{3.50}$$

It is interesting to observe that, although the coupling matrices $\mathbf{V}^{(i)}$ and $\mathbf{U}^{(i)}$ of the individual elements are frequency independent, the overall coupling matrices $\mathbf{V}^{(i)}(\omega)$ and $\mathbf{U}^{(i)}(\omega)$ are frequency dependent.

In simulations, the per-segment $\tau^{(i)}$ can be constant, while the coupling matrices $\mathbf{V}^{(i)}$ and $\mathbf{U}^{(i)}$ vary in each simulation round, turning the accumulated group delay τ into a random variable. Variable τ follows a bilateral Maxwell distribution, and its modulus, $|\tau|$, follows a Maxwell distribution

$$f(|\tau|) = \sqrt{\frac{2}{\pi}} \tau^2 e^{-\tau^2/2}. \tag{3.51}$$

The mean and the variance of the Maxwell distribution are given by $\overline{|\tau|} = 2\sqrt{2/\pi}$ and $\sigma_{|\tau|}^2 = 3 - 8/\pi$. It can be shown that the standard deviation of τ, σ_τ^2, scales with the number of sections N as [13]

$$\sigma_\tau = \sqrt{N} \sigma_{\tau^{(i)}}. \tag{3.52}$$

The PMD of optical fibers is typically specified in terms of its mean DGD $\overline{|\tau|}$. It can be shown that the standard deviation of the bilateral Maxwell σ_τ and the mean of the Maxwell distribution $\overline{|\tau|}$ are related by $\sigma_\tau = \sqrt{3\pi/8}\,\overline{|\tau|}$. Therefore,

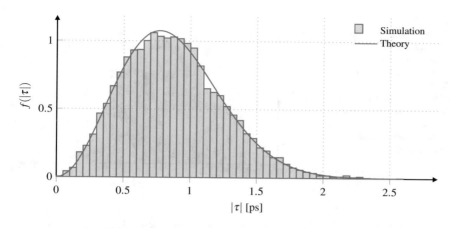

Fig. 3.8 Empirical and theoretical distributions of the accumulated DGD $\overline{|\tau|}$ of a 300-km fiber link with mean DGD 0.05 ps/$\sqrt{\text{km}}$. The theoretical curve (solid line) is the Maxwell distribution with mean $\overline{|\tau|} = 0.05 \times \sqrt{300} = 0.866$ ps. The empirical distribution (bars) are obtained through the Monte Carlo simulation (10,000 rounds) of 200 birefringent elements with $\tau^{(i)} = \sqrt{3\pi/(8 \times 200)} \times 0.866 = 0.0664$ ps

simulating an accumulated mean DGD of $\overline{|\tau|}$ requires a per-section DGD $\tau^{(i)} = \sqrt{3\pi/(8N)}\,\overline{|\tau|}$.

As an example, typical optical fibers have a DGD specification of 0.05 ps/$\sqrt{\text{km}}$. Simulating a 300-km optical link would correspond to an accumulated DGD of $\overline{|\tau|} = 0.05 \times \sqrt{300} = 0.866$ ps. Therefore, simulating this fiber link with 200 sections would require a per-span DGD of $\tau^{(i)} = \sqrt{3\pi/(8 \times 200)} \times 0.866 = 0.0664$ ps. The theoretical and empirical distributions of $\overline{|\tau|}$ corresponding to this case are shown in Fig. 3.8.

Polarization effects have a stochastic nature, as a result of temperature variations and mechanical stress caused, for example, by wind in aerial cables, or vibrations in a bridge crossed by the optical fiber cable. For this reason, polarization effects are compensated by linear adaptive filters in optical systems with coherent detection. The architecture of these filters will be studied in Chap. 5.

3.4 Attenuation

Attenuation reduces the power of an optical signal propagating through the fiber. It is directly related to the system reach, as receivers have a minimum level of power for which they can recover the transmitted data satisfactorily. Attenuation occurs due to several phenomena, such as material absorption, which is intrinsic to the material used in the production of the fiber and its impurities; scattering of light; imperfections in the waveguide geometry; and curvatures in the fiber. Because of

Fig. 3.9 Attenuation profile of an optical fiber. Optical communications operate around three transmission windows centered around 850, 1310, and 1550 nm. Coherent systems used for long-distance transmission operate mainly in the third transmission window, which also exhibits the lowest loss

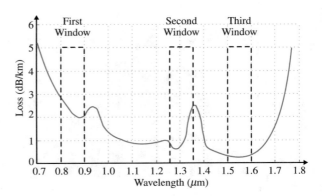

attenuation, the input power $P_{i-\text{dBm}}$ and the output power $P_{o-\text{dBm}}$ of an optical fiber, in dBm, are related by

$$P_{o-\text{dBm}} = P_{i-\text{dBm}} - \alpha_{\text{dB/km}} L_{\text{km}}, \tag{3.53}$$

where L_{km} is the fiber length in km and $\alpha_{\text{dB/km}}$ is the attenuation coefficient given in dB/km. It is also possible to describe the same relationship in terms of linear power units, with P_o and P_i expressed in mW (or W)

$$P_{o-\text{mW}} = P_{i-\text{mW}} e^{-\alpha_{\text{Np/km}} L_{\text{km}}}, \tag{3.54}$$

where $\alpha_{\text{Np/km}}$ is the attenuation coefficient given in Np/km. The attenuation coefficient α in dB/km and in Np/km are related by

$$\alpha_{\text{dB/km}} = (10 \log_{10} e) \, \alpha_{\text{Np/km}} = 4.34 \alpha_{\text{Np/km}}. \tag{3.55}$$

The attenuation profile of an optical fiber shows the dependence of attenuation on the wavelength. It has three main low-attenuation windows that are typically used for optical communications, around 850, 1310, and 1550 nm, as illustrated in Fig. 3.9. The steep increase of attenuation in wavelengths shorter than 850 nm is mainly caused by Rayleigh scattering due to random fluctuations in the silica refractive index. On the other side of the spectrum, the high increase in attenuation in wavelengths longer than 1600 nm appears because of absorption in the infrared region. Another dominant phenomenon is absorption around 1390 nm caused by residual OH$^-$ ions in silica. This peak divides the second and third windows, centered at 1310 (with $\alpha_{\text{dB/km}} \approx 0.3$ dB/km) nm and 1550 nm (with $\alpha_{\text{dB/km}} \approx 0.2$ dB/km). New processes of optical fiber manufacturing manage to practically eliminate the OH$^-$ peak (also called water peak), enabling a wider transmission bandwidth. The third window, around 1550 nm, exhibits an attenuation of approximately 0.2 dB/km, which is considerably lower than that for the other two, and is therefore widely used for long-distance transmission.

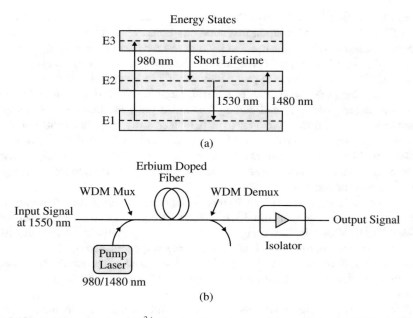

Fig. 3.10 (**a**) Energy levels of Er^{3+} ions in silica glass used for optical amplification. Optical amplification around the 1550 nm band requires high ion population in E2. Amplification can be achieved in two ways. The first way is to pump ions from E1 to E3 at 980 nm. As the lifetime of ions in E3 is short ($\tau_{32} = 10\,\mu s$) [16], ions quickly transition to E2 through the process of spontaneous emission. The second way is to pump ions directly to E2 at 1480 nm, where the ion lifetime is longer ($\tau_{21} = 10\,ms$). (**b**) EDFA architecture. A pump laser at 980 or 1480 nm is coupled to the signal through a WDM mux device. The coupled signals are launched into an erbium-doped fiber. Pump and signal are then decoupled by a WDM demux. An isolator prevents backscattering of light into the amplifier

3.5 Optical Amplification

The simplest solution to attenuation would be to increase the power with which optical signals are launched into the fiber. However, the indiscriminate increase in power strengthens nonlinear fiber effects that degrade the system performance. The alternative solution adopted in long-distance optical systems is the periodic insertion of optical amplifiers along the link. Among existing solutions, the erbium-doped fiber amplifier (EDFA) is the most widely used.

EDFAs operate typically in the C-band (1530–1565 nm), but L-band (1565–1625 nm) designs are also available commercially. Optical amplification in EDFAs is based on the process of stimulated emission (just as lasers). The choice of the erbium ion (Er^{3+}) as dopant is due to the fact that the difference between two of its energy levels corresponds to wavelengths close to 1550 nm, which is exactly the lowest attenuation band in the fiber. Figure 3.10a shows the energy levels of Er^{3+} ions on silica involved in the EDFA amplification process. It is desired to obtain a large population in E2 to favor the stimulated emission process in the transition from E1. Such population inversion can be achieved in two ways. The first way is to pump ions from E1 to E3 at 980 nm. As the lifetime of ions in E3 is short ($\tau_{32} = 10\,\mu s$) [16], ions quickly transition to E2 through the process of spontaneous emission. The second way is to pump ions directly to E2 at 1480 nm, where the ion lifetime is longer ($\tau_{21} = 10\,ms$). In general, pumping at 980 nm is more efficient, attaining larger gains [17] and adding less noise [18]. In addition, 980-pumping also facilitates pump multiplexing, as it is further away from the 1550-nm signal channels. On the other hand, as the fiber has a lower attenuation at 1480 nm than at 980 nm, 1480-nm pumping is recommended in specific applications involving remote pumping as in hostile areas, where optical pumps propagate along the signal over the transmission fiber.

The typical architecture of an EDFA is depicted in Fig. 3.10b. A pump laser operating at the wavelengths of 980 or 1480 nm is coupled to the doped fiber using a WDM multiplexer. Both the signal and the pump propagate through the EDF, leading to signal amplification. Finally, the pump is separated again from the signal by a WDM demultiplexer. An isolator is usually placed at the amplifier output to prevent backscattering of light into the amplifier because of the high output power levels.

The purpose of optical amplifiers is to imprint optical gain to input signals. Naturally, this gain is limited, constrained by the pump signal and the properties of the gain medium. For high input powers, the amplifier gain is reduced, leading the amplifier into saturation. This phenomenon can be analyzed through the expression of the large-signal amplifier gain [11]

$$G = G_0 e^{\left(-\frac{G-1}{G}\frac{P_{out}}{P_s}\right)} = G_0 e^{\left(-(G-1)\frac{P_{in}}{P_s}\right)}, \tag{3.56}$$

where $P_{out} = G P_{in}$ is the signal output power, P_{in} is the signal input power, G is the amplifier gain, G_0 is the small-signal amplifier gain, and P_s is the saturation power. This is a transcendental equation for G, i.e., it does not have a closed-form solution. Clearly, for low input powers P_{in}, the amplifier applies gain G_0. However, for higher input powers, the gain decreases, and the amplifier reaches saturation. This effect is observed in Fig. 3.11, which shows G as a function of P_{in} for $G_0 = 20\,dB$ and several P_s values.

During propagation through the doped fiber, the launched optical signal triggers stimulated emissions that, in turn, contribute to signal amplification. However, spontaneous photon emissions are also amplified, giving rise to amplified spontaneous emission (ASE) noise. The amplifier performance in terms of noise is quantified by its noise figure F. In general, the noise figure of a device is defined as [19]

$$F = \frac{\text{SNR}_{in}}{\text{SNR}_{out}}, \tag{3.57}$$

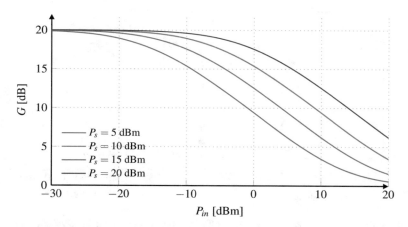

Fig. 3.11 Amplifier gain G as a function of the input power P_{in}, for $G_0 = 20\,\text{dB}$ and different values of saturation power P_s. At low input power levels, the small-signal gain G_0 is attained. For higher values of P_{in}, the amplifier gain G decreases as a result of amplifier saturation

where SNR_{in} and SNR_{out} correspond to the input and output signal-to-noise ratios (SNR). However, the definition of F is not as simple as it looks. Optical communications are impaired by several noise sources, both optical and electric, so the noise figure requires a more detailed definition. Usually, the noise figure of optical amplifiers is defined considering the electric SNR at an ideal hypothetical receiver accounting for shot noise at the device input, and the beat of ASE noise with the signal at the output [20]. For most of the G values found in practice, the amplifier noise figure F can be reduced to a simple expression related to the amplifier spontaneous emission factor n_{sp} [11]

$$F \approx 2n_{sp}. \tag{3.58}$$

Factor n_{sp}, in turn, is given by

$$n_{sp} = N_2/(N_2 - N_1), \tag{3.59}$$

where N_1 and N_2 are the atomic populations for the ground and excited states.

The ASE noise generated in amplifiers depends on the loss and gain coefficients of the doped fiber. Figure 3.12a shows the absorption and gain coefficients of an EDFA pumped at 980 nm [21]. As expected, gain surpasses absorption in most of the C-band. The balance of EDF gain and loss gives rise to an uneven gain spectrum, as shown in Fig. 3.12b, which depicts the ASE spectrum (that coincides with the gain profile) for a sample optical amplifier pumped at 980 nm. This spectral gain unbalance is usually compensated by gain-flattening filters (GFFs) specially designed to yield a flat spectrum at the EDFA output. In spite of the high

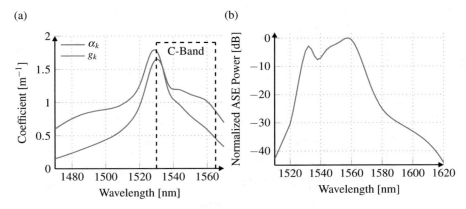

Fig. 3.12 (**a**) Absorption α_k and gain g_k coefficients for an EDF pumped at 980 nm [21]. Gain surpasses absorption in the vicinity of the C-band. (**b**) Normalized ASE spectrum for a sample EDFA pumped at 980 nm

effectiveness of GFFs, the EDFA gain profile still depends on several variables, such as amplifier gain and input power, and residual unbalances can still appear.

Optical amplifiers can be classified into three types according to their functionality, as depicted in Fig. 3.13. The amplifier positioned just after the transmitter is called a *booster*. It has the purpose of providing very high power output levels (e.g., 20 dBm) before launching the signals into the transmission fiber. As its input power is relatively high, its gain is relatively low and, therefore, its noise figure is not of primary importance (e.g., 6 dB). The *pre-amplifier* is positioned immediately before the optical receiver, and its purpose is to raise the signal power per channel to a level higher than the receiver sensitivity. Therefore, it must have a high gain, a relatively low output power (e.g., 13 dBm), and a low noise figure (e.g., 5 dB). The amplifiers positioned along the link are called *in-line* amplifiers, and their purpose is to increase the optical signal powers from very low values at their inputs to very high values at their outputs. Therefore, it must combine the best properties of pre-amplifiers and boosters, i.e., high gain, high output power, and low noise figure. This is sometimes achieved by a two-stage architecture that has a pre-amplifier in the first stage and booster in the second stage. The segment of fiber between two amplifiers is called a span. In typical optical terrestrial systems, the span length varies in the range of 80–100 km. In submarine transoceanic systems, these numbers can be considerably lower, reaching 40–60 km.

ASE noise is usually modeled as AWGN. Although the output WDM spectrum is not perfectly flat, the noise spectrum can be considered flat over the signal bandwidth. ASE noise accumulates along the link due to the traversal of several stages of amplification, progressively degrading the signal quality. The typical measure of noise accumulation in optical fiber systems is the optical signal-to-noise ratio (OSNR), defined as the ratio between the average power of the optical signal

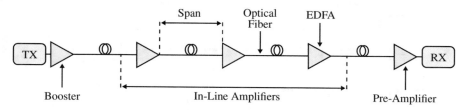

Fig. 3.13 Amplified optical system. The booster is located at the transmit site and operates with high input power and high output power. The pre-amplifier is located at the receive site and operates with low input power and low output power. The in-line amplifier is located at intermediate amplifier sites and operates with low input power and high output power. In-line amplifiers are often designed in two stages, the first stage being a pre-amplifier and the second stage a booster

per channel, P_{signal}, and the power of the ASE noise, P_{ASE}, measured in a reference bandwidth B_{ref} of 12.5 GHz

$$\text{OSNR} = \frac{P_{signal}}{P_{ASE}} = \frac{P_{signal}}{2G_{ASE}B_{ref}}, \tag{3.60}$$

where G_{ASE} is the single-polarization noise power spectrum density. The factor of two in the denominator of (3.60) accounts for the two polarization components. The single-polarization power spectrum density of the ASE noise generated by an EDFA, that compensates for the losses of a single span of length L_s, G_{ASE}^{SS}, is given by Essiambre [22]

$$G_{ASE}^{SS} = (e^{\alpha L_s} - 1)h\nu_s n_{sp}, \tag{3.61}$$

where h is Planck's constant and ν_s is the operating frequency. The power attenuation factor α must be expressed in Np/km.

Considering a link with N_s spans, where each in-line and pre-amplifier EDFA compensates for propagation losses in the preceding span, the OSNR in the end of the link is given by

$$\text{OSNR} = \frac{P_{signal}}{2N_s G_{ASE}^{SS} B_{ref}} = \frac{P_{signal}}{2N_s(e^{\alpha L_s} - 1)h\nu_s n_{sp} B_{ref}}. \tag{3.62}$$

Finally, knowing that $F \approx 2n_{sp}$, the expression for the OSNR becomes

$$\text{OSNR} \approx \frac{P_{signal}}{F N_s(e^{\alpha L_s} - 1)h\nu_s B_{ref}}. \tag{3.63}$$

Figure 3.14 shows the OSNR versus link length curve for a system configuration with $P_{signal} = 0\,\text{dBm}$, $L_s = 80\,\text{km}$, $\alpha_{dB} = 0.2\,\text{dB/km}$, and amplifiers with $F = 5\,\text{dB}$.

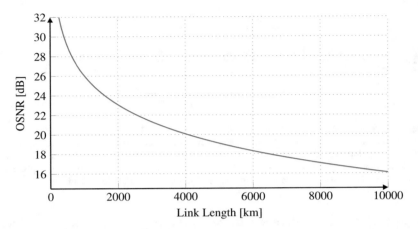

Fig. 3.14 OSNR *versus* link length ($N_s L_s$) profile, for $P_{\text{signal}} = 0\,\text{dBm}$, $L_s = 80\,\text{km}$, $\alpha_{\text{dB}} = 0.2\,\text{dB/km}$, and amplifiers with $F = 5\,\text{dB}$

The OSNR can be related to the SNR by assuming minimum-bandwidth Nyquist-shaped pulses [22], and transmission and reception in one polarization orientation [22]

$$\text{SNR} = \frac{E_s}{N_0} = \frac{P_{\text{signal}}}{G_{\text{ASE}} R_s} = \frac{2 B_{\text{ref}}}{R_s} \frac{P_{\text{signal}}}{P_{\text{ASE}}} = \frac{2 B_{\text{ref}}}{R_s} \text{OSNR}, \tag{3.64}$$

where the signal bandwidth is assumed to be the symbol rate R_s, and the noise power spectrum density in one polarization G_{ASE}. Note that, in the derivation of (3.64), we denote $E_s = P_{\text{signal}}$ as the signal power after a matched receiver, and N_0 as the noise power after the matched receiver [23].

In systems with polarization multiplexing, P_{signal} is equally divided between the two polarization orientations and, therefore,

$$\text{SNR} = \frac{B_{\text{ref}}}{R_s} \text{OSNR} \tag{3.65}$$

Figure 3.15 shows the BER (calculated according to (2.18)) *versus* OSNR curve, for the 16-QAM and QPSK modulation formats considering polarization multiplexing and a symbol rate of 28 GBd. Using the relation between BER and OSNR in Fig. 3.15, and OSNR and link length in Fig. 3.14, it is possible to estimate BER as a function of the link length, as shown in Fig. 3.16. For a given symbol rate, 16-QAM conveys twice the bit rate compared with QPSK. On the other hand, for a given BER, the maximum link length is much shorter than for QPSK. Currently, flexible transceivers allow the user to select the desired modulation format from a set of options. Indeed, there are several other parameters that can be configured, which, for now, will not be studied, such as symbol rate and code rate. Choosing the right

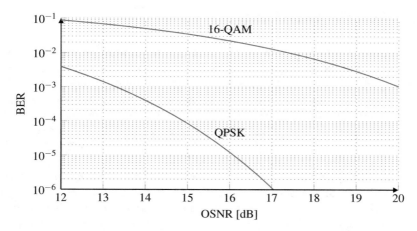

Fig. 3.15 BER *versus* OSNR curve for the 16-QAM and QPSK modulation formats considering polarization multiplexing and a symbol rate of 28 GBd. This configuration yields a total bit rate of 112 Gb/s for QPSK and 224 Gb/s for 16-QAM

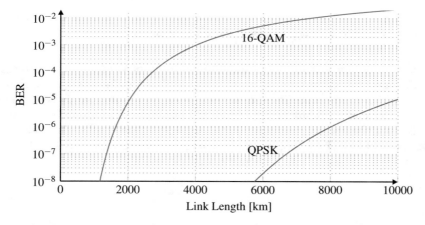

Fig. 3.16 BER *versus* link length curve for the 16-QAM and QPSK modulation formats considering polarization multiplexing and a symbol rate of 28 GBd. The optical link parameters are $P_{\text{signal}} = 0\,\text{dBm}$, $L_s = 80\,\text{km}$, $\alpha_{\text{dB}} = 0.2\,\text{dB/km}$, and amplifiers with $F = 5\,\text{dB}$. This configuration yields a total bit rate of 112 Gb/s for QPSK and 224 Gb/s for 16-QAM

modulation format is a complex task. The choices are simplified when it is assumed that the transmission bandwidth is limited and the coding scheme is fixed. In this case, for a given symbol rate, the modulation format is then chosen to provide the highest bit rate for a given link length.

3.6   Nonlinear Interference

The field profile of the fundamental mode supported in single-mode fibers causes the largest portion of energy associated with the electromagnetic field to be confined within the core. Consequently, given the small core size, the spatial power density is relatively high. This makes optical fiber propagation nonlinear in certain configurations, particularly at the first kilometers of fiber after optical WDM amplification. The nonlinear effects generated in optical fibers are subdivided into two groups. The first group consists of effects caused by inelastic scattering, in which stimulated Brillouin scattering (SBS) and stimulated Raman scattering (SRS) are the most important examples. These effects cause the optical signal power delivered to the fiber to be transferred to scattered waves, reducing the received signal power or causing inter-channel interference. In coherent optical WDM systems, the transparent reach is essentially limited by the second group of nonlinearities, which comprises the effects originated from the dependence of the refractive index with the intensity of the applied field, the so-called Kerr effect.

3.6.1   Effective Length and Effective Area

The impact of nonlinear effects on fiber depends on the power density, or brightness, on the fiber cross-sectional area. As the field amplitude of the propagating signals varies with the radial coordinate, calculating the impact of nonlinear effects on the fiber is challenging. Alternatively, the concept of effective area is used. In a single-mode fiber, the field amplitude profile follows Bessel functions. The mathematical evaluation of field effects can be simplified by approximating the Bessel field profile by a Gaussian function with parameter w_0, called spot size

$$|E(r)| \approx E_0 e^{-r^2/w_0^2}, \tag{3.66}$$

where E_0 is the field amplitude at the center of the fiber. In a single-mode fiber, the ratio of w_0 by the core radius a can be approximated by a function of the normalized frequency v as [11]

$$w_0/a \approx 0.65 + 1.619v^{-3/2} + 2.879v^{-6}. \tag{3.67}$$

The approximation in (3.67) has practical importance and is particularly accurate for $v \approx 2$ [11]. While the field profile follows a Gaussian-like function with parameter w_0, calculations involving nonlinearities assume a step function with width w_0, as shown in Fig. 3.17 [11]. The effective area A_{eff} parameter is then defined as the area determined by this equivalent profile

$$A_{\text{eff}} = \pi w_0^2. \tag{3.68}$$

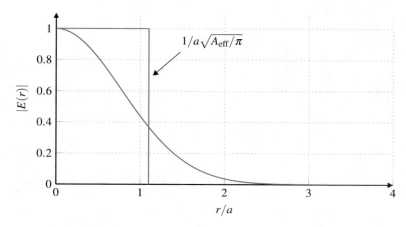

Fig. 3.17 Gaussian field amplitude $|E(r)|$ as a function of the radial coordinate r, normalized by the core radius a, for $v = 2.4$ (blue), and the equivalent field amplitude used in calculations involving the fiber effective area (red)

Just as assuming a constant power profile over the cross-sectional area simplifies calculations, assuming a constant power profile in the longitudinal direction is also a usual practice. The optical power decreases exponentially in the longitudinal direction according to the fiber attenuation constant. As nonlinear effects are generated at high power profiles, nonlinearities are generated only in the first kilometers after optical amplification in long enough spans, whereas propagation is linear in the last kilometers before reception. The effective length L_{eff} is calculated assuming that the integral of the field power over the longitudinal direction z, for a constant power profile, is the same as for the exponential profile (see Fig. 3.18)

$$P_0 L_{\mathrm{eff}} = \int_0^L P_0 e^{-\alpha z} dz, \tag{3.69}$$

where L is the fiber length, α is the power attenuation constant (in Np/km), and P_0 is the power at the fiber input. Thus, L_{eff} is given by

$$L_{\mathrm{eff}} = \frac{1 - e^{-\alpha L}}{\alpha}. \tag{3.70}$$

For long fibers, $L_{\mathrm{eff}} \approx 1/\alpha$. For example, for $\alpha_{\mathrm{dB/km}} = 0.2\,\mathrm{dB/km}$, we have $\alpha_{\mathrm{Np/km}} = 0.046\,\mathrm{Np/km}$. In this case, for a long fiber, $L_{\mathrm{eff}} \approx 1/\alpha_{\mathrm{Np/km}} = 21\,\mathrm{km}$. This result indicates that, in a long SSMF, practically all nonlinear effects occur in the first 20 km of fiber.

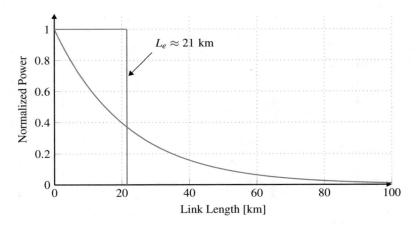

Fig. 3.18 Normalized power as a function of link length, for $\alpha = 0.2\,\text{dB/km}$ (blue), and normalized power assumed in calculations using the effective length and a constant power profile (red). In this case, $L_e \approx 21\,\text{km}$

3.6.2   Inelastic Scattering

In inelastic scattering, a photon interacts with the molecular structure of the material, in such a way as to generate a second photon with less energy, and a phonon, which is the oscillatory motion of atoms or molecules in a specific frequency. The difference in frequency of the incident and scattered optical waves is called Stokes shift. In SBS, photons are scattered in the backward direction, generating a Stokes shift in the order of 10 GHz with a bandwidth below 100 MHz. The SBS amplification gain has a Lorentzian profile [11], as shown in Fig. 3.19a for a Stokes shift of 11.15 GHz and a bandwidth of 20 MHz. The SBS effect is significant in systems with lower data rates and consequently narrow spectral widths. Its practical effect is to limit the power that reaches the receiver, as a large portion can be backscattered toward the transmitter. However, even in these low-data-rate systems SBS is not necessarily a limiting effect, as its efficiency may be strongly suppressed by artificially dithering the phase of the transmission carrier. In coherent optical systems with high data rates, SBS is not a limiting effect.

SRS also scatters light, transferring energy from short-wavelength channels to long-wavelength channels. However, unlike SBS, which scatters light in the backward direction only, SRS scatters light in both the forward and backward directions. The SRS gain has a peak around 13 THz with a spectral range of 20–30 THz, as shown in Fig. 3.19b for an SSMF. In WDM systems, SRS causes a tilt in spectrum that affects transmission in the entire WDM band. As it is a cumulative effect, its impact can be mitigated by equalizing the WDM power spectrum periodically. The relevance of SRS in coherent optical systems, however, stands out in the use of amplifiers based on the SRS effect. Raman amplifiers are normally positioned before EDFA pre-amplifiers in very long spans to provide additional gains. In its

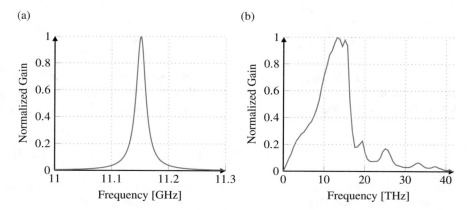

Fig. 3.19 (**a**) Normalized gain spectrum generated by SBS. The spectrum has a Lorentzian shape calculated with central frequency of 11.15 GHz and 3-dB bandwidth of 20 MHz. (**b**) Normalized gain spectrum generated by SRS on an SSMF fiber pumped at 1430 nm. The SRS gain spectrum has a peak at 13 THz with a bandwidth of 20–30 THz

most traditional application, a continuous-wave pump signal is launched into the counter-propagating direction. The pump signal can combine multiple wavelengths in order to ensure a minimum gain flatness for the amplification process.

3.6.3 Kerr Effect

In Sect. 3.1 we learned that silica glasses respond to applied fields with a linear polarization vector $\mathbf{P} = \mathbf{P_l}$ that generates important fiber effects such as CD. However, this is an approximation and, in addition to the linear response, in silica glasses the polarization vector $\mathbf{P}$ also has a nonlinear contribution $\mathbf{P_{nl}}$

$$\mathbf{P} = \mathbf{P_l} + \mathbf{P_{nl}}. \tag{3.71}$$

In silica fibers, the dominant nonlinear term is given by the third-order nonlinearity, given by Agrawal [24]

$$\mathbf{P_{nl}} = \epsilon_0 \chi^{(3)} \mathbf{EEE}, \tag{3.72}$$

where $\chi^{(3)}$ is the fourth-rank tensor that operates on the input electric field vectors. Notice that, as we consider nonlinear contributions, we cannot operate in this moment in frequency domain with $\tilde{\mathbf{E}}$ but rather operate in time domain with $\mathbf{E} = \mathbf{E}(\mathbf{r}, t)$. Assuming an instantaneous nonlinear response, and avoiding tensors, the expression for component i (x, y, or z) of $\mathbf{P_{nl}}$ can be given by New [25]

$$P_{nl_i} = \epsilon_0 \sum_{j=1}^{3} \sum_{k=1}^{3} \sum_{l=1}^{3} \chi_{ijkl}^{(3)} E_j E_k E_l \quad (i, j, k, l = x, y, z), \tag{3.73}$$

where $\chi_{ijkl}^{(3)}$ are the third-order nonlinear coefficients. For the sake of simplicity, we now assume the field $\mathbf{E}$ polarized along a single direction, having therefore a single component E_i. In this case, $\mathbf{E}$ and $\mathbf{P}$ can be represented by scalars. Then, (3.73) can be rewritten as [16]

$$P_{nl} = \epsilon_0 \chi^{(3)} E^3(\mathbf{r}, t). \tag{3.74}$$

where $\chi^{(3)} = \chi_{iiii}^{(3)}$.

The third-order nonlinear polarization term gives rise to an intensity-dependent refractive index, an effect that was first observed by J. Kerr in 1875 [26]. To understand this relationship, let us assume the transmission of a polarized monochromatic wave. Then, $E(\mathbf{r}, t) = E_0 \cos(\omega_0 t - \beta_0 z)$. In this case, the nonlinear polarization is given by Ramaswami et al. [16]

$$P_{nl} = \epsilon_0 \chi^{(3)} E_0^3 \cos^3(\omega_0 t - \beta_0 z) \tag{3.75}$$

$$= \epsilon_0 \chi^{(3)} E_0^3 \left(\frac{3}{4} \cos(\omega_0 t - \beta_0 z) + \frac{1}{4} \cos(3\omega_0 t - 3\beta_0 z) \right) \tag{3.76}$$

Because of phase matching conditions, term with $3\omega_0$ does not consistently propagate. The relationship between ω and β for supported modes is defined by the modal map in Fig. 3.4. As this relationship is nonlinear, although the (ω_0, β_0) pair is a valid solution, the $(3\omega_0, 3\beta_0)$ pair is not. Therefore,

$$P_{nl} \approx \frac{3}{4} \epsilon_0 \chi^{(3)} E_0^2 E(\mathbf{r}, t) = \epsilon_0 \epsilon_{nl} E(\mathbf{r}, t), \tag{3.77}$$

where the nonlinear dielectric constant is given by

$$\epsilon_{nl} = \frac{3}{4} \chi^{(3)} E_0^2. \tag{3.78}$$

Neglecting losses, the refractive index is given by $n = \sqrt{\epsilon}$. In the linear regime, n is independent of the signal intensity, as

$$n_l = \sqrt{1 + \chi^{(1)}}. \tag{3.79}$$

In the nonlinear regime, however, there is an additional term dependent on the signal intensity

$$n_{nl} = \sqrt{1 + \chi^{(1)} + \frac{3}{4}\chi^{(3)}E_0^2} = \sqrt{n_l^2 + \frac{3}{4}\chi^{(3)}E_0^2} = n_l\sqrt{1 + \frac{3}{4n_l^2}\chi^{(3)}E_0^2}.$$

(3.80)

As the nonlinear term is very small, the nonlinear refractive index can be approximated by

$$n_{nl} \approx n_l + \frac{3}{8n_l}\chi^{(3)}E_0^2.$$

(3.81)

Equation (3.81) shows that the refractive index in silica is modulated by the intensity of the electric field. The equation indicates that, in amplitude-modulated signals, the phase of the optical signal is also modulated, causing spectral broadening, as shown in Fig. 3.20a. This is the effect of self-phase modulation (SPM). In optical fiber communications, (3.81) is also commonly expressed as

$$n_{nl} \approx n_l + \bar{n}_2 E_0^2,$$

(3.82)

where $\bar{n}_2 = \frac{3}{8n_l}\chi^{(3)}$ is the *nonlinear index coefficient* [27]. Equation (3.82) can be applied to the core or the cladding of a single-mode fiber.

It is also possible to determine the approximate nonlinear contribution to the mode propagation constant as [24]

$$\beta_{nl} \approx \beta_l + \gamma E_0^2,$$

(3.83)

where $\gamma = 2\pi\bar{n}_2/(\lambda A_{\text{eff}})$ is the fiber *nonlinear parameter*. Equation (3.83) reveals the increase in the nonlinear phase shift with the inverse of the effective area.

If, instead of a single monochromatic signal, two sinusoidal signals in frequencies ω_1 and ω_2 and with amplitudes E_1 and E_2 are present, the nonlinear polarization becomes [16]

$$P_{nl} = \epsilon_0\chi^{(3)}\left[E_1\cos(\omega_1 t - \beta_1 z) + E_2\cos(\omega_2 t - \beta_2 z)\right]^3$$

(3.84)

$$= \epsilon_0\chi^{(3)}\left[\left(\frac{3E_1^3}{4} + \frac{3E_2^2 E_1}{2}\right)\cos(\omega_1 t - \beta_1 z)\right.$$

$$+ \left.\left(\frac{3E_2^3}{4} + \frac{3E_1^2 E_2}{2}\right)\cos(\omega_2 t - \beta_2 z)\right]$$

$$+ \epsilon_0\chi^{(3)}\left(\frac{3E_1^2 E_2}{4}\right)\cos((2\omega_1 - \omega_2)t - (2\beta_1 - \beta_2)z)$$

$$+ \epsilon_0\chi^{(3)}\left(\frac{3E_2^2 E_1}{4}\right)\cos((2\omega_2 - \omega_1)t - (2\beta_2 - \beta_1)z)$$

$$+ \ldots$$

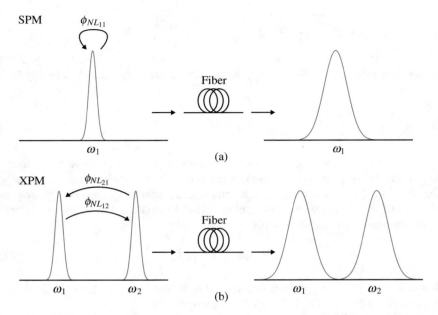

Fig. 3.20 (**a**) In SPM, amplitude variations of a signal generate a pattern-dependent nonlinear phase shift $\phi_{NL_{11}}$ on itself, causing spectral broadening and impairing transmission. (**b**) In XPM, amplitude variations of a signal in frequency ω_1 (or ω_2) generate a pattern-dependent nonlinear phase shift $\phi_{NL_{12}}$ (or $\phi_{NL_{21}}$) on a second signal of frequency ω_2 (or ω_1), causing spectral broadening and impairing transmission

where the last terms are omitted for not supporting the phase matching condition. The terms with $(3E_i^3/4)$ correspond to SPM, where intensity variations in channel i with (i=1,2) induce phase variations on itself. Terms with $(3E_i^2 E_j/2)$ correspond to the case where intensity variations in channel i (i=1,2) generate phase variations in channel $j \neq i$, as illustrated in Fig. 3.20b. This is the effect of cross-phase modulation (XPM). There are also terms in (3.84) with new components being generated at frequencies $2\omega_1 - \omega_2$ and $2\omega_2 - \omega_1$. Notice that these contributions do satisfy phase matching conditions, as these new wavelengths and propagation constants are relatively close to the original ones. This effect is known as degenerate four-wave mixing (FWM).

Adding a third wavelength component to $E(\mathbf{r}, t)$ generates a large number of contributions. Some of them satisfy the phase matching condition, and others will not. In addition to the SPM and XPM contributions, new wavelengths are generated at frequencies

$$\omega_{ijk} = \omega_i + \omega_j - \omega_k \quad k \neq i, j. \tag{3.85}$$

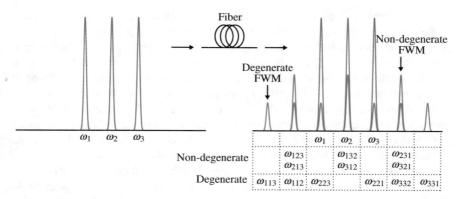

Fig. 3.21 In FWM, three signals in frequencies ω_i, ω_j, and ω_k generate a fourth signal in frequency $\omega_{ijk} = \omega_i + \omega_j - \omega_k$, with $k \neq i, j$. If $i = j$, the products are called degenerate, while if $i \neq j$, they are called non-degenerate

Note that $i = j$ corresponds to the degenerate FWM case. FWM received this name as signals in three frequencies interact to generate a fourth one, as indicated in Fig. 3.21.

In the recent history of optical fiber communications, the Kerr effect has been studied in the given settings in accordance with its outcomes as SPM, XPM, or FWM [24], depending on the number of mixing frequencies. The study of these effects enabled the dimensioning of several generations of optical fiber systems based on on–off keying modulation and dispersion management. However, modern communication systems with Nyquist pulse shaping and electronic dispersion compensation have properties that favor other approaches for the study of nonlinearities. These approaches are based on the observation that, in long-distance dispersion-uncompensated systems, the nonlinear interference can be modeled as AWGN for the purpose of system modeling [6].

3.6.4 The Gaussian Noise Model

The Gaussian noise (GN) model is based on the observation that, in long dispersion-uncompensated systems, the nonlinear Kerr interference behaves as AWGN. The question becomes then how to calculate the variance of this nonlinear contribution for a wavelength channel. The solution to this problem can be found in a four-wave mixing model for nonlinear interference. The third-order nonlinearity raises the propagating field to the third power, generating new interference components. As we have seen in the previous section, by considering the mixing of three wavelengths, all nonlinear contributions, including SPM, XPM, and FWM, are accounted for. Unlike the early studies on nonlinearity modeling, where FWM was studied based on the mixing of different WDM channels, the GN model discretizes the entire spectrum into narrowband subcarriers, including individual channels, as if they were orthogonal frequency division multiplexing (OFDM) signals. Such approach has been adopted in as early as 1993 by Splett et al. [28], and later in 2003 by Louchet et al. [29]. After some time in background, the study of FWM-based models was resumed by investigations on the performance of OFDM in optical transmission systems [30, 31]. However, the use of these models in the design of practical systems would only be consolidated a few years later. With the popularization of coherent optical communication systems, the inline compensation of CD was eliminated, altering the statistical behavior of nonlinear interference. As we have mentioned before, it was shown that, in long-haul systems on SSMFs and without CD compensators, nonlinear interference behaves approximately as additive Gaussian noise [6]. This empirical observation paved the way for several later works that developed the foundations of what is known today as the Gaussian noise model, or simply GN model [7, 8, 32, 33].

The Gaussian noise model is based on a reference expression for the power spectrum density of nonlinear interference, given by Poggiolini [7]

$$G_{\text{NLI}}(f) = \frac{16}{27} \gamma^2 L_{\text{eff}}^2 \int_{-\infty}^{\infty} \int_{-\infty}^{\infty} G_{\text{WDM}}(f_1) G_{\text{WDM}}(f_2) G_{\text{WDM}}(f_1 + f_2 - f) \cdot$$

$$\rho(f_1, f_2, f) \cdot \chi(f_1, f_2, f) \, df_2 \, df_1, \tag{3.86}$$

where γ is the fiber nonlinear coefficient, L_{eff} is the fiber effective length, and G_{WDM} is the power spectrum density of the WDM propagating signal. The FWM efficiency factor ρ is given by

$$\rho(f_1, f_2, f) = \left| \frac{1 - e^{[j4\pi^2|\beta_2|(f_1-f)(f_2-f)-\alpha]L_s}}{\alpha - j4\pi^2|\beta_2|(f_1-f)(f_2-f)} \right|^2 \cdot L_{\text{eff}}^{-2}, \tag{3.87}$$

where L_s is the span length and α is the power attenuation coefficient[2] in Np/km. The phase array factor χ determines the degree of coherence in nonlinear interference accumulation along the link and has an expression equivalent to that obtained for the field of an antenna array [31]

$$\chi(f_1, f_2, f) = \frac{\sin^2 \left(2N_s \pi^2 (f_1 - f)(f_2 - f)\beta_2 L_s\right)}{\sin^2 \left(2\pi^2 (f_1 - f)(f_2 - f)\beta_2 L_s\right)}, \tag{3.88}$$

where N_s is the number of spans. Equation (3.86) is valid under the following conditions: (1) the system does not have dispersion compensation modules; (2) transmission is carried out with polarization multiplexing; (3) all fiber spans have

---

[2]Note that most publications about the GN model use the amplitude attenuation coefficient, which is half of the power attenuation coefficient.

the same length L_s; 3) amplifiers are not saturated, and their gain equals the span losses; and (4) the third-order dispersion parameter (β_3) is zero.

The computation of Eq. (3.86) is usually nontrivial, particularly because of the definition of integration domains. A conservative approximation to (3.86) can be made by assuming that the power spectrum density of the propagating signals is flat over the entire WDM spectrum, neglecting the nonzero roll-off factor of the transmitted signal shape, as well as guard bands between channels. Thus, assuming minimum-bandwidth Nyquist-shaped transmission, and zero guard bands, $G_{WDM}(f) = G_{WDM}$ within the WDM spectrum, from $-B_{WDM}/2$ to $B_{WDM}/2$.

A simple expression can be obtained by evaluating (3.86) at $f = 0$ and $N_s = 1$ (single-span assumption). Setting $f = 0$ corresponds to a worst-case value, as the central frequency of the WDM spectrum receives nonlinear interference from close neighbors of both sides. The dependence with $N_s > 1$ can be reintroduced with simple scaling rules [7]. Under these assumptions, the power spectrum density at $f = 0$ and generated in a single span, $G_{NLI,SS}$, is given by Poggiolini [7]

$$
G_{NLI,SS} = \frac{16}{27}\gamma^2 G_{WDM}^2 \int_{-B_{WDM}/2}^{B_{WDM}/2} \int_{-B_{WDM}/2}^{B_{WDM}/2} G_{WDM}(f_1 + f_2) \cdot
$$

$$
\left| \frac{1 - e^{j4\pi^2|\beta_2|f_1 f_2 L_s - \alpha L_s}}{\alpha - j4\pi^2|\beta_2|f_1 f_2} \right|^2 \cdot df_1 \, df_2. \tag{3.89}
$$

Equation (3.89) can be further approximated as [7]

$$
G_{NLI,SS} \approx \frac{8}{27}\gamma^2 G_{WDM}^3 L_{eff}^2 \frac{\operatorname{asinh}\left(\frac{\pi^2}{2}\beta_2 L_{eff,a} B_{WDM}^2\right)}{\pi \beta_2 L_{eff,a}}, \tag{3.90}
$$

where $L_{eff,a} = 1/\alpha$ is the asymptotic effective length.

The modeling of nonlinearities can be simplified by assuming that the nonlinear interference power scales linearly with the number of spans, that is, the nonlinear power P_{NLI} in the reference bandwidth B_{ref} is given by Poggiolini [7] and Poggiolini et al. [8]

$$
P_{NLI} = N_s G_{NLI,SS} B_{ref}. \tag{3.91}
$$

Under these simplifying conditions, it is possible to redefine the OSNR, taking into account the noise generated by nonlinear interference

$$
OSNR' = \frac{P_{TX}}{P_{ASE} + P_{NLI}}. \tag{3.92}
$$

So far, we have assumed a constant transmit power per channel P_{TX}, but we have not discussed how to select P_{TX}. As we already know, P_{TX} cannot be increased indefinitely to improve OSNR', as the nonlinear component in the denominator of (3.92) increases faster than P_{TX} in the numerator.

As the Kerr effect is a third-order nonlinearity, the nonlinear interference scales with the third power of the signal power

$$P_{NLI} = N_s \eta_{SS} P_{TX}^3, \tag{3.93}$$

where η_{SS} is a nonlinearity efficiency factor for a single span. By inspection of (3.90) and (3.92), and making $G_{WDM} = P_{TX}/R_s$ (minimum-bandwidth Nyquist pulse shaping), where R_s is the symbol rate, we obtain

$$\eta_{SS} \approx \frac{8}{27} \gamma^2 L_{eff}^2 \frac{\text{asinh}\left(\frac{\pi^2}{2} \beta_2 L_{eff,a} B_{WDM}^2\right)}{\pi \beta_2 L_{eff,a}} \frac{B_{ref}}{R_s^3}. \tag{3.94}$$

Figure 3.22 shows the nonlinear interference efficiency ($N_s \eta_{SS}$) in dB, calculated according to (3.94), considering $\gamma = 1.3$ 1/W/km, $L_{eff} = L_{eff,a} = 21$ km, $B_{WDM} = 4$ THz, $D = 17$ ps/nm/km, $\lambda = 1550$ nm, $R_s = 50$ GHz, and $B_{ref} = 12.5$ GHz.

Equation (3.92) can then be rewritten as

$$OSNR' = \frac{P_{TX}}{P_{ASE} + N_s \eta_{SS} P_{TX}^3}. \tag{3.95}$$

Differentiating (3.95) with respect to P_{TX}, and setting the result to zero, gives an optimum transmit power $P_{TX,OPT}$

$$P_{TX,OPT} = \left(\frac{P_{ASE}}{2 N_s \eta_{SS}}\right)^{\frac{1}{3}}. \tag{3.96}$$

Plugging (3.96) into (3.92), we find that

$$OSNR'_{MAX} = \frac{P_{TX,OPT}}{P_{ASE} + 0.5 P_{ASE}}. \tag{3.97}$$

Thus, for small transmit powers, ONSR' is low, dominated by ASE noise. For high transmit powers, OSNR' again decreases, dominated by nonlinear interference. The OSNR is maximized in a quasi-linear regime, in which the nonlinear interference power is half of the ASE noise power, as shown in Fig. 3.23. Indeed, the design of optical transmission systems is the art of optimizing the launch power and finding the proper balance between the ASE and nonlinear interference powers.

Finally, for a given required OSNR' (OSNR'$_{req}$), the maximum reach without regeneration, called transparent reach L, is given by

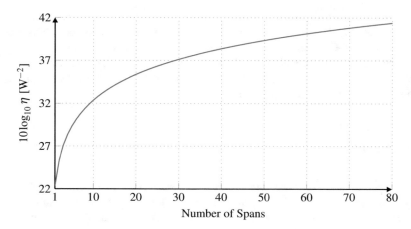

Fig. 3.22 Nonlinear interference efficiency ($N_s \eta_{SS}$), calculated according to (3.94), considering $\gamma = 1.3\,\text{W}^{-1}/\text{km}$, $L_{\text{eff}} = L_{\text{eff,a}} = 21\,\text{km}$, $B_{\text{WDM}} = 4\,\text{THz}$, $D = 17\,\text{ps/nm/km}$, $\lambda = 1550\,\text{nm}$, $R_s = 50\,\text{GHz}$, and $B_{\text{ref}} = 12.5\,\text{GHz}$

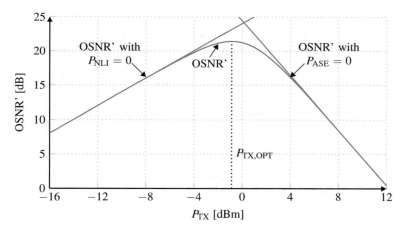

Fig. 3.23 OSNR' *versus* P_{TX} curve (solid line) and optimum transmit power (dotted line), for $L_s = 80\,\text{km}$, $N_s = 20$, $\alpha_{\text{dB}} = 0.2\,\text{dB/km}$, $\nu_s = 193.1\,\text{THz}$, F=5 dB, and $\eta_{ss} = 180\,\text{W}^{-2}$. OSNR' is maximized at $P_{TX,OPT}$

$$L = N_s L_s = \frac{P_{\text{Tx,OPT}}}{\text{OSNR'}_{\text{req}}\left[F(e^{\alpha L_s} - 1)h\nu B_{\text{ref}} + \eta_{SS}P_{\text{TX,OPT}}^3\right]} L_s. \qquad (3.98)$$

The GN model is very broad and has many variants. The variant presented in this section was chosen for its simplicity and illustrative value. In any case, the presented methodology provides a valid order of magnitude for nonlinear interference for the design of WDM systems. Improvements to the original GN model have been published by several groups [33, 34]. These improvements show that the nonlinear interference also depends on other parameters that have not been taken into account by the original GN model, such as modulation format and number of channels considering a fully occupied WDM spectrum. Even in the face of these more accurate models, the GN model still presents itself as a valid approach, providing conservative and low complexity predictions.

3.7   Problems

1. Calculate the maximum core radius a of a step-index fiber with core refractive index $n_1 = 1.45$ and fractional index change at the core–cladding interface $\Delta = (n_1 - n_2)/n_1 = 0.003$, for single-mode condition at $\lambda = 1550$ nm.
2. For a step-index fiber with core radius $a = 4\,\mu$m, core refractive index $n_1 = 1.45$ and fractional index change at the core–cladding interface $\Delta = (n_1 - n_2)/n_1 = 0.003$, calculate the minimum wavelength λ for single-mode operation.
3. For a step-index fiber with core radius $a = 4\,\mu$m and core refractive index $n_1 = 1.45$, calculate the maximum fractional index change at the core–cladding interface $\Delta = (n_1 - n_2)/n_1$, for single-mode operation at 1550 nm. Use $(n_1 + n_2)/n_1 \approx 2$.
4. Suppose you wish to design a space-division multiplexing (SDM) multimode step-index fiber supporting 12 parallel transmission flows. The 12 flows correspond to 6 spatial modes, having 2 polarization modes each. The operating wavelength is 1550 nm. If the core refractive index is $n_1 = 1.45$ and the fractional index change at the core–cladding interface $\Delta = (n_1 - n_2)/n_1 = 0.003$, which is the range of core radii of the designed fiber? In your solution, consider all modes, including degenerate polarization (x and y) and spatial (a and b) modes. Use the table with the roots for the Bessel functions of the first kind $J_l(x)$.
5. An uncompensated 2000-km long-distance fiber link using an SSMF with $D = 17$ ps/nm/km operates at a symbol rate of 50 GBaud with minimum-bandwidth Nyquist pulse shaping at 1550 nm. How many symbols approximately are the channel delay spread?
6. The table below shows the properties of SSMFs and DSFs

Parameter	SSMF	DSF
λ_0	1.3 μm	1.5 μm
Effective area	85 μm^2	40 μm^2

where λ_0 is the wavelength for which the GVD parameter $D = 0$. Based on the parameters above, and knowing that nonlinear effects are enhanced at small effective areas, discuss the advantages and disadvantages of DSFs compared with SSMFs.

7. Generate a pulse-shaped waveform with 10 QPSK symbols by convolving the generated symbol sequence with a root-raised cosine (RRC) filter impulse response. Use an RRC filter with roll-off factor $\beta^{RC} = 0.1$ and a span of 20 symbols, assuming 16 Sa/Symbol. Assuming a symbol rate of 50 GBaud, plot the modulus of the transmitted symbol sequence and of the received symbol sequence in the optical domain after a 1-km SSMF link with $D = 17$ ps/nm/km. Repeat the problem with a 10-km SSMF link.

8. Suppose a 300-km optical fiber link with DGD parameter $\overline{|\tau|} = 0.05$ ps/$\sqrt{\text{km}}$. Calculate analytically the mean and standard deviation of the accumulated DGD in this link.

9. Simulate 1000 realizations of a 300-km SSMF fiber link with 200 sections and DGD parameter $\overline{|\tau|} = 0.05$ ps/$\sqrt{\text{km}}$. Using the group delay operator, estimate the DGD $|\tau|$ generated in each realization. Calculate the mean and the standard deviation of $|\tau|$ from the simulated data, and compare them with the expected theoretical values.

10. Generate a pulse-shaped waveform with 10 QPSK symbols as described in Problem 7. Assume a symbol rate of 50 GBaud. Simulate the PMD channel of a 1-km SSMF link with 200 sections and average DGD parameter $\overline{|\tau|} = 0.5$ ps/$\sqrt{\text{km}}$. Plot the modulus of the optical transmitted symbol sequence in one polarization, and of the received symbol sequence, for two realizations of the PMD channel. To calculate the modulus of the received symbol sequence, consider the square root of the sum of the squared modulus of each of the two polarization components at the output of the PMD channel. Repeat the problem for a 300-km SSMF link.

11. An unamplified 50-km SSMF link operating at 1550 nm uses a 10-Gb/s transceiver with −23-dBm sensitivity. Calculate the minimum transmit power per channel in mW for proper system operation. Repeat the problem if the system operates at 1310 nm.

12. An unamplified SSMF link operating at 1550 nm uses a 10-Gb/s transceiver with −1-dBm transmitter power and −23-dBm sensitivity. Calculate the maximum link length for proper system operation. Repeat the problem if the system operates at 1310 nm.

13. An EDFA has the following operation conditions. (1) It amplifies a signal from 16 μW to 10 mW. (2) It amplifies a signal from 54 μW to 20 mW. Use approximations to estimate the saturation power P_s of this EDFA.

14. The output saturation power P_s^{out} is defined as the output power at which the amplifier gain is reduced by 3 dB. Calculate the approximate output saturation power P_s^{out} as a function of the saturation power P_s.

15. Consider an amplified optical system with the below parameters:
Span length L_s: 80 km
Number of spans N_s: 125
Transmit power P_{signal}: 0 dBm
Attenuation constant $\alpha_{dB/km}$: 0.2 dB/km

Operating frequency ν: 193.1 THz
Planck's constant h: $6.62 \cdot 10^{-34}$ J $\cdot$ s
Amplifier noise figure: 5 dB
Polarization Multiplexing

(a) Plot the theoretic curve of OSNR in dB *versus* link distance in km.
(b) Plot the theoretic curves of BER *versus* OSNR in dB, for the QPSK and 16-QAM modulation formats, at the symbol rates of 28 GBaud and 50 GBaud.
(c) Plot the theoretic curves of BER *versus* link distance, for the QPSK and 16-QAM modulation formats, at the symbol rates of 28 GBaud and 50 GBaud.
(d) Repeat (c) considering a 3-dB OSNR margin (subtract 3 dB from the calculated OSNR).
(e) Calculate the system reaches for the configurations in (c), assuming that the transceiver tolerates a BER $= 10^{-3}$.
(f) Repeat (e) considering a 3-dB OSNR margin (subtract 3 dB from the calculated OSNR).

16. Plot simulated curves of BER *versus* OSNR in dB, for Nyquist-shaped QPSK and 16-QAM signals, at the symbol rates of 28 GBaud and 50 GBaud, assuming single polarization transmission. For pulse shaping, use an RRC filter with $\beta^{RC} = 0.1$ and a span of 20 symbols, assuming 16 Sa/Symbol. Consider OSNRs in the interval from 7 dB to 17 dB. Include in your simulator MZM modulation and a matched filter at the receiver. Compare the results with the theoretical curves.

17. Repeat Problem 16, including in your simulator dispersion corresponding to a 3-km SSMF ($D = 17$ ps/nm/km and $\lambda = 1550$ nm). Consider OSNRs in the interval from 7 dB to 22 dB. At the symbol rate of 28 GBaud, determine the OSNR penalty caused by dispersion at BER $= 10^{-3}$.

18. Calculate the effective area of an SSMF and of an DSF at 1550 nm and using $n_1 = 1.45$. The SSMF has a fractional index change at the core–cladding interface $\Delta = (n_1 - n_2)/n_1 = 0.003$ and a core radius $a = 4.1\,\mu$m. For the DSF, use $\Delta = 0.0075$ and a core radius $a = 2.3\,\mu$m.

19. Calculate the pump wavelength of a Raman amplifier with peak amplification at the center of the C-band (1530–1565 nm), and at the center of the L-band (1565–1625 nm).

20. Calculate the frequencies of degenerate FWM products generated by signals at frequencies at 193.1 and 193.2 THz.

21. Calculate the frequencies of degenerate and non-degenerate FWM products generated by three equally spaced signal channels, with frequencies ω_1, $\omega_1 + \Delta\omega$, and $\omega_1 + 2\Delta\omega$. Calculate the number of products that interfere with each of the transmitted signals.

22. Using a simplified GN model, calculate the nonlinear interference efficiency η_{SS} for a WDM transmission system with the following parameters: $\gamma = 1.3$ 1/W/km, $L_{eff} = L_{eff,a} = 21$ km, $B_{WDM} = 4$ THz, $D = 17$ ps/nm/km, $\lambda = 1550$ nm, $R_s = 50$ GHz, and $B_{ref} = 12.5$ GHz.

23. Consider the same parameters of the previous problem and, additionally, that the amplifier noise figure $F = 5\,\text{dB}$ and that the span length $L_s = 80\,\text{km}$. Calculate the optimum transmit power $P_{\text{TX,OPT}}$.
24. Consider the same parameters as in the previous problem. Assuming a required $\text{OSNR}'_{\text{req}} = 13\,\text{dB}$, calculate the transparent reach at the optimum transmit power $P_{\text{TX,OPT}}$. Repeat the problem for transmit powers $1\,\text{dBm}$ below and $1\,\text{dBm}$ above $P_{\text{TX,OPT}}$.

3.8 Matlab/Octave Functions

3.8.1 Functions for Section 3.2

Matlab/Octave Code 3.1 Insertion of CD in signals.

```
function [AOutput] = CDInsertion(AInput,SpS,Rs,D,CLambda,L,NPol)
%%%%%%%%%%%%%%%%%%%%%%%%%%%%%%%%%%%%%%%%%%%%%%%%%%%%%%%%%%%%%%%%%%%%%%%%%%%%
% CDInsertion [AOutput] = CDInsertion(AInput,SpS,Rs,D,CLambda,L,NPol)    %
%                                                                        %
%   This function simulates the insertion of chromatic dispersion (CD) in %
% the signal 'AInput'.                                                    %
%                                                                        %
% Input:                                                                  %
%  AInput   = Input Signal. For transmission in single pol. orientation  %
%             'AInput' must be a column vector. For transmission with pol. %
%             multiplexing, 'AInput' must be a matrix with two columns,   %
%             where each column corresponds to the signal of a pol.       %
%             orientation (V and H pol. orientations);                    %
%  SpS      = Number of samples per symbol in the input signal 'AInput';  %
%  Rs       = Symbol rate in [symbols/s];                                 %
%  D        = Dispersion parameter in [ps/(nm x km)]                      %
%  CLambda  = Central lambda in [m]                                       %
%  L        = Fiber length in [m]                                         %
%  NPol     = Number of polarizations used;                               %
%                                                                        %
% Output:                                                                 %
%   AOutput = Output signal after CD insertion. 'AOutput' is arranged in  %
%             columns in the same way as 'AInput';                        %
%                                                                        %
% This function is part of the book Digital Coherent Optical Systems;    %
% Darli A. A. Mello and Fabio A. Barbosa;                                %
%%%%%%%%%%%%%%%%%%%%%%%%%%%%%%%%%%%%%%%%%%%%%%%%%%%%%%%%%%%%%%%%%%%%%%%%%%%%

    % Constants:
    c = 299792458;

    % Dispersion:
    D = D*1e-12/(1e-9*1e3);

    % Frequency vector:
    w = 2*pi*(-1/2:1/size(AInput,1):1/2-1/size(AInput,1)).'*SpS*Rs;

    % Calculating the CD frequency response:
    G = exp(1i*((D*CLambda^2)/(4*pi*c))*L*w.^2);

    % Inserting CD to the transmitted signal:
    AOutput(:,1) = ifft(ifftshift(G.*fftshift(fft(AInput(:,1)))));
```

```
    % In the case of pol-mux:
    if NPol == 2
    % Inserting CD to the transmitted signal:
        AOutput(:,2) = ifft(ifftshift(G.*fftshift(fft(AInput(:,2)))));
    end
end
```

3.8.2   Functions for Section 3.3

Matlab/Octave Code 3.2   Insertion of PMD in signals.

```
function [EOutput,varargout] = PMDInsertion(EInput,DGDSpec,L,N,Rs,SpS,...
    EvalGroupDelay)
%%%%%%%%%%%%%%%%%%%%%%%%%%%%%%%%%%%%%%%%%%%%%%%%%%%%%%%%%%%%%%%%%%%%%%%%%%%%
% PMDINSERTION [EOutput,varargout] = PMDInsertion(EInput,DGDSpec,L,N,Rs,  %
%                                    SpS,EvalGroupDelay)                  %
%                                                                        %
%    This function simulates the PMD insertion in pol. multiplexed signals %
% ('EInput'). The PMD model considers a cascade of 'N' sections, a mean  %
% DGD of 'DGDSpec', and a fiber of length 'L'. This function also        %
% estimates the group delay of one realization of the 'PMD channel' when %
% the flag 'EvalGroupDelay' is set to true.                              %
%                                                                        %
% Input:                                                                 %
%   EInput        = Input signal. 'EInput' must be a matrix with two     %
%                   columns, where each column has the signal of a pol.  %
%                   orientation (V and H pol. orientations);             %
%   DGDSpec       = Mean DGD in [ps/(km)^(1/2)];                         %
%   N             = Number of sections of the PMD model;                 %
%   L             = Fiber length [m];                                    %
%   Rs            = Symbol rate in [symbols/s];                          %
%   SpS           = Number of samples per symbol in the input signal     %
%                   'EInput';                                            %
%   EvalGroupDelay = Variable to enable (true) or disable (false) the    %
%                   group delay estimation of one realization of the 'PMD%
%                   channel'. To estimate the distribution of the        %
%                   accumulated DGD, the group delay of several          %
%                   realizations of the 'PMD channel' must be done.      %
%                                                                        %
%   *Note: If this function is only used to obtain empirically the       %
%     distribution of the accumulated DGD, 'EInput', 'Rs', and 'SpS' can be%
%     defined as: 'EInput = [0 0]' ; 'Rs = 0' ; 'SpS = 0';              %
%   *Note: When EvalGroupDelay = false, the function has only one output;%
%                                                                        %
% Output:                                                                %
%   EOutput    = Output signal after PMD insertion. 'EOutput' is arranged %
%                in columns in the same way as 'EInput';                %
%   varargout = Group delay in [ps], when 'EvalGroupDelay = true';      %
%                                                                        %
% This function is part of the book Digital Coherent Optical Systems;    %
% Darli A. A. Mello and Fabio A. Barbosa;                                %
%%%%%%%%%%%%%%%%%%%%%%%%%%%%%%%%%%%%%%%%%%%%%%%%%%%%%%%%%%%%%%%%%%%%%%%%%%%%

    % Standard deviation of the Maxwellian distribution:
    SDTau = sqrt(3*pi/8)*DGDSpec;

    % DGD per section (it is equal to the standard deviation per section):
    Tau = (SDTau*sqrt(L*1e-3)/sqrt(N))*1e-12;

    % Frequency vector:
```

```
w = 2*pi*fftshift(-1/2:1/size(EInput,1):1/2-1/size(EInput,1)).'...
    *SpS*Rs;

% Random unitary matrices V and U that describes mode coupling:
for i = 1:N
    [V(:,:,i),~,U(:,:,i)] = svd(randn(2) + 1i*randn(2));
end

% Estimation of the group delay (GD operator):
if EvalGroupDelay
    % Frequencies to consider when evaluating the group delay:
    wGD = [1 1.1];

    % Obtaining the transfer matrix H:
    for k = 1:numel(wGD)
        % Auxiliary matrix:
        HAux = 1;

        % Delay matrix (Lambda):
        Lambda = [exp(1i*wGD(k)*Tau/2) 0; 0 exp(-1i*wGD(k)*Tau/2)];

        % Transfer function of the i-th section and matrix H:
        for i = 1:N
            Hi = V(:,:,i)*Lambda*U(:,:,i)' ; HAux = HAux*Hi;
        end

        % Matrix H for frequency indicated by k:
        H(:,:,k) = HAux;
    end

    % Obtaining the eigenvalues (num. differentiating H):
    HDiff       = (H(:,:,2)-H(:,:,1))/(wGD(2)-wGD(1));
    Eigenvalues = eig(1i*HDiff/(H(:,:,2)));

    % Group delays in ps:
    GroupDelay   = abs(real(2*Eigenvalues(1)*1e12));
    varargout{1} = GroupDelay;
end

% Signals in frequency domain:
Freq_E_V = fft(EInput(:,1)) ; Freq_E_H = fft(EInput(:,2));

for i = 1:N
    % Hermitian of matrix U:
    UHermitian = U(:,:,i)';

    % Rotating the signals according to U:
    E_1 = UHermitian(1,1)*Freq_E_V + UHermitian(1,2)*Freq_E_H;
    E_2 = UHermitian(2,1)*Freq_E_V + UHermitian(2,2)*Freq_E_H;

    % Applying DGD:
    E_1 = exp(1i*w*Tau/2).*E_1 ; E_2 = exp(-1i*w*Tau/2).*E_2;

    % Rotating the signals according to V:
    Freq_E_V = V(1,1,i)*E_1 + V(1,2,i)*E_2;
    Freq_E_H = V(2,1,i)*E_1 + V(2,2,i)*E_2;
end

% Signals in time domain:
EOutput(:,1) = ifft(Freq_E_V) ; EOutput(:,2) = ifft(Freq_E_H);
end
```

3.8.3 Functions for Section 3.5

Matlab/Octave Code 3.3 AWGN generation considering a given OSNR in dB.

```
function [r] = NoiseInsertion_OSNR(x,OSNRdB,SpS,NPol,Rs,BRef)
%%%%%%%%%%%%%%%%%%%%%%%%%%%%%%%%%%%%%%%%%%%%%%%%%%%%%%%%%%%%%%%%%%%%%%%%%%%
% NOISEINSERTION_OSNR [r] = NoiseInsertion_OSNR(x,OSNRdB,SpS,NPol,Rs,Bref)%
%                                                                        %
%    This function inserts additive white Gaussian noise (AWGN) in the   %
% transmitted signal 'x' (in single- or dual-polarization orientations) so%
% that an OSNR (in dB) of 'OSNRdB' is achieved.                          %
%                                                                        %
% Input:                                                                 %
%   x      = Transmitted signal. For transmission in single pol.         %
%            orientation, 'x' must be a column vector. For transmission   %
%            with pol. multiplexing, 'x' must be a matrix with two column,%
%            where each column vector corresponds to the signal of a pol. %
%            orientation (V and H pol. orientations);                    %
%   OSNRdB = OSNR in dB;                                                  %
%   SpS    = Number of samples per symbol in the input signal 'x';       %
%   NPol   = Number of polarizations to be used.                         %
%   Rs     = Symbol rate in symbols/second;                              %
%   BRef   = Reference bandwidth (in Hz) for OSNR measurement;           %
%                                                                        %
% Output:                                                                %
%   r = Signal after noise insertion. 'r' is arranged in columns in the  %
%       same way as 'x';                                                 %
%                                                                        %
% This function is part of the book Digital Coherent Optical Systems;    %
% Darli A. A. Mello and Fabio A. Barbosa;                                %
%%%%%%%%%%%%%%%%%%%%%%%%%%%%%%%%%%%%%%%%%%%%%%%%%%%%%%%%%%%%%%%%%%%%%%%%%%%

    % OSNR/SNR in linear scale:
    OSNRLin = 10^(OSNRdB/10);
    SNRLin  = ((2*BRef)/(NPol*Rs))*OSNRLin;

    % AWGN standard deviation:
    StdDev = sqrt(mean(abs(x(:,1)).^2)*SpS/(2*SNRLin));

    % AWGN generation:
    n = StdDev*randn(length(x(:,1)),1)+1i*StdDev*randn(length(x(:,1)),1);

    % Inserting noise to the signal:
    r(:,1) = x(:,1) + n;

    % In the case of pol-mux:
    if NPol == 2
      % AWGN standard deviation:
      StdDev = sqrt(mean(abs(x(:,2)).^2)*SpS/(2*SNRLin));

      % AWGN generation:
      n = StdDev*randn(length(x(:,2)),1)+1i*StdDev*randn(length(x(:,2)),1);

      % Inserting noise to the signal:
      r(:,2) = x(:,2) + n;
    end
end
```

References

1. E. Ip, J.M. Kahn, Compensation of dispersion and nonlinear impairments using digital backpropagation. J. Lightwave Technol. **26**(20), 3416–3425 (2008)
2. X. Li, X. Chen, G. Goldfarb, E. Mateo, I. Kim, F. Yaman, G. Li, Electronic post-compensation of WDM transmission impairments using coherent detection and digital signal processing. Opt. Express **16**(2), 880–888 (2008)
3. R. Dar, P.J. Winzer, Nonlinear interference mitigation: methods and potential gain. J. Lightwave Technol. **35**(4), 903–930 (2017)
4. T. Lima, V. Rozental, A. Barreto, D. Mello, Network-efficient superchannel transmission by the multichannel compensation of nonlinearities, in *Proceedings of Optical Fiber Communication Conference (OFC)* (Optical Society of America, Washington, 2014), p. Th1E.2
5. E.F. Mateo, G. Li, Compensation of interchannel nonlinearities using enhanced coupled equations for digital backward propagation. Appl. Opt. **48**(25), F6–F10 (2009)
6. P. Poggiolini, A. Carena, V. Curri, G. Bosco, F. Forghieri, Analytical modeling of nonlinear propagation in uncompensated optical transmission links. IEEE Photon. Technol. Lett. **23**(11), 742–744 (2011)
7. P. Poggiolini, The GN model of non-linear propagation in uncompensated coherent optical systems. J. Lightwave Technol. **30**(24), 3857–3879 (2012)
8. P. Poggiolini, G. Bosco, A. Carena, V. Curri, Y. Jiang, F. Forghieri, The GN-model of fiber non-linear propagation and its applications. J. Lightwave Technol. **32**(4), 694–721 (2014)
9. S. Bottacchi, *Multi-Gigabit Transmission over Multimode Optical Fibre: Theory and Design Methods for 10GbE Systems* (Wiley, Chichester, 2006)
10. D. Gloge, Weakly guiding fibers. Appl. Opt. **10**(10), 2252–2258 (1971)
11. G.P. Agrawal, *Fiber-Optic Communication Systems*, 4th ed. (Wiley, Hoboken, 2010)
12. M. Shtaif, A. Mecozzi, Modelling of polarization mode dispersion in optical communications systems, in *Polarization Mode Dispersion*, ed. by A. Galtarossa, C.R. Menyuk (Springer, New York, 2005), ch. 2
13. K. Ho, J.M. Kahn, Statistics of group delays in multimode fiber with strong mode coupling. J. Lightwave Technol. **29**(21), 3119–3128 (2011)
14. K.-P. Ho, J.M. Kahn, Mode-dependent loss and gain: statistics and effect on mode-division multiplexing. Opt. Express **19**(17), 16612–16635 (2011)
15. D.A.A. Mello, H. Srinivas, K. Choutagunta, J.M. Kahn, Impact of polarization- and mode-dependent gain on the capacity of ultra-long-haul systems. J. Lightwave Technol. **38**(2), 303–318 (2020)
16. R. Ramaswami, K. Sivarajan, G. Sasaki, *Optical Networks: A Practical Perspective*, 3rd ed. (Morgan Kaufmann Publishers, Francisco, 2009)
17. M. Yamada, M. Shimizu, T. Takeshita, M. Okayasu, M. Horiguchi, S. Uehara, E. Sugita, Er^{3+}-doped fiber amplifier pumped by 0.98 μm laser diodes. IEEE Photon. Technol. Lett. **1**(12), 422–424 (1989)
18. M. Yamada, M. Shimizu, M. Okayasu, T. Takeshita, M. Horiguchi, Y. Tachikawa, E. Sugita, Noise characteristics of Er^{3+}-doped fiber amplifiers pumped by 0.98 and 1.48 μm laser diodes. IEEE Photon. Technol. Lett. **2**(3), 205–207 (1990)
19. H.T. Friis, Noise figures of radio receivers. Proc. IRE **32**(7), 419–422 (1944)
20. H.A. Haus, The noise figure of optical amplifiers. IEEE Photon. Technol. Lett. **10**(11), 1602–1604 (1998)
21. J.K. Perin, J.M. Kahn, J.D. Downie, J. Hurley, K. Bennett, Importance of amplifier physics in maximizing the capacity of submarine links. J. Lightwave Technol. **37**(9), 2076–2085 (2019)
22. R.J. Essiambre, G. Kramer, P.J. Winzer, G.J. Foschini, B. Goebel, Capacity limits of optical fiber networks. J. Lightwave Technol. **28**(4), 662–701 (2010)
23. J.R. Barry, D.G. Messerschmitt, E.A. Lee, *Digital Communication*, 3rd ed. (Kluwer Academic Publishers, Norwell, 2004)
24. G. Agrawal, *Nonlinear Fiber Optics*, 3rd ed. (Academic Press, San Diego, 2001)

25. G. New, *Introduction to Nonlinear Optics* (Cambridge University Press, Cambridge, 2011)
26. J. Kerr, A new relation between electricity and light: dielectrified media birefringent. Lond. Edinb Dublin Philos. Mag. J. Sci. **50**(332), 337–348 (1875)
27. T. Kato, Y. Suetsugu, M. Takagi, E. Sasaoka, M. Nishimura, Measurement of the nonlinear refractive index in optical fiber by the cross-phase-modulation method with depolarized pump light. Opt. Lett. **20**(9), 988–990 (1995)
28. A. Splett, C. Kurtzke, K. Petermann, Ultimate transmission capacity of amplified optical fiber communication systems taking into account fiber nonlinearities, in *Proceedings of European Conference on Optical Communication (ECOC), September* (1993)
29. H. Louchet, A. Hodzic, K. Petermann, Analytical model for the performance evaluation of DWDM transmission systems. IEEE Photon. Technol. Lett. **15**(9), 1219–1221 (2003)
30. A.J. Lowery, S. Wang, M. Premaratne, Calculation of power limit due to fiber nonlinearity in optical OFDM systems. Opt. Express **15**(20), 13282–13287 (2007)
31. M. Nazarathy, J. Khurgin, R. Weidenfeld, Y. Meiman, P. Cho, R. Noe, I. Shpantzer, V. Karagodsky, Phased-array cancellation of nonlinear FWM in coherent OFDM dispersive multi-span links. Opt. Express **16**(20), 15777–15810 (2008)
32. A. Carena, V. Curri, G. Bosco, P. Poggiolini, F. Forghieri, Modeling of the impact of nonlinear propagation effects in uncompensated optical coherent transmission links. J. Lightwave Technol. **30**(10), 1524–1539 (2012)
33. P. Poggiolini, Modeling of non-linear propagation in uncompensated coherent systems. in *Proceedings of Optical Fiber Communication Conference* (Optical Society of America, Washington, 2013), p. OTh3G.1
34. R. Dar, M. Feder, A. Mecozzi, M. Shtaif, Properties of nonlinear noise in long, dispersion-uncompensated fiber links. Opt. Express **21**(22), 25685–25699 (2013)

Chapter 4
The Receiver Front-End, Orthogonalization, and Deskew

A typical digital coherent optical receiver is composed of two major subsystems, as shown in Fig. 4.1. The first subsystem, indicated in grey, is the receiver front-end, consisting of an optical front-end, transimpedance amplifiers (TIAs), and analog-to-digital converters (ADCs). The optical front-end of a dual-polarization optical receiver produces four electric outputs $i_{IV}(t)$, $i_{QV}(t)$, $i_{IH}(t)$, and $i_{QH}(t)$, corresponding to the in-phase and quadrature components of two orthogonal polarization orientations, denoted here as vertical (V) and horizontal (H). These currents are converted to voltage by a set TIAs, and digitized by a set of ADCs, generating the digital signals $r_{IV}[n]$, $r_{QV}[n]$, $r_{IH}[n]$, and $r_{QH}[n]$ [1, 2]. The resulting signals are sent to the second subsystem, indicated in blue, consisting of a chain of digital signal processing (DSP) algorithms.

The first DSP block is that of *deskew*, responsible for compensating for possible temporal mismatches in the alignment of the four received components [3, 4]. The parameters of the deskew block are typically static and factory-characterized. After the temporal alignment carried out by the deskew block, the in-phase and quadrature components of the received signals can be combined, producing two complex signals, corresponding to the V and H polarization orientations. The next DSP block is that of *orthogonalization*, which compensates for the mismatches on the receiver front-end that are not related to temporal misalignments, such as slightly unbalanced photodetectors or power splitters [5]. Then, two static filters perform *chromatic dispersion (CD) compensation* [6]. The vast majority of current optical coherent systems operate without any optical CD compensation, resulting in high intersymbol interference (ISI) to be dealt with at the receiver. Therefore, these filters are the longest of the coherent receiver, and their length can reach thousands of taps. Fortunately, CD is a practically time-invariant effect, and its compensation does not require temporal adaptability. The following block is that of symbol synchronization, also known as *clock recovery* [7]. Although it is placed after CD compensation in Fig. 4.1, it can also be implemented in other positions of the DSP chain, such as after adaptive equalization, or even within the CD

© Springer Nature Switzerland AG 2021
D. A. de Arruda Mello, F. A. Barbosa, *Digital Coherent Optical Systems*,
Optical Networks, https://doi.org/10.1007/978-3-030-66541-8_4

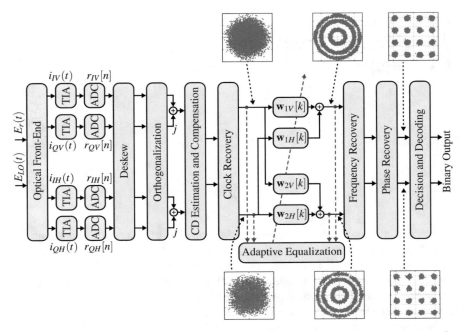

Fig. 4.1 Digital coherent receiver. The coherent receiver is composed of two main blocks, the receiver front-end (in gray) and the chain of DSP algorithms (in blue). The receiver front-end beats the received polarization-multiplexed optical signal $E_r(t)$ with the local oscillator $E_{LO}(t)$, and produces four digital signals corresponding to the in-phase and quadrature components of the vertical and horizonal polarization orientations. The chain of DSP algorithms processes these digital signals and attempts to recover the transmitted sequence

compensation block. Clock recovery detects possible mismatches between the symbol rate and the ADC sampling rate, and implements a control loop that corrects this mismatch digitally using time-varying interpolators, or actuating directly on the ADC reference oscillator.

Following clock recovery, an *adaptive equalizer* with two inputs and two outputs interconnected in butterfly structure compensates for polarization effects and other eventual residual linear distortions. The adaptive equalizer also separates the polarization-multiplexed signals [8]. Figure 4.1 shows the constellations before and after adaptive equalization. The equalizer inputs are linear combinations of the transmitted symbols, resulting in constellations with a circular shape. At the equalizer outputs the signals are separated. The resulting constellations have multiple rings, corresponding to the transmitted constellation affected by phase rotations.

After the adaptive equalizer, a *frequency recovery* block compensates for the frequency mismatch between transmission and local oscillator lasers [9]. Phase noise effects are mitigated by a *phase recovery* block [10]. After phase and frequency recovery, the shape of the transmitted constellation is recovered. Finally the decision block retrieves the decoded bit sequence. This chapter presents the receiver front-end, and discusses the algorithms used to correct its imperfections using deskew and orthogonalization methods.

4.1 The Receiver Front-end

In optical receivers, the conversion of the optical signal into electric currents is performed by the photodetector. Photodetectors are optoelectronic devices that generate an electric current $I_p(t)$ that is proportional to the squared modulus of the optical input field $E_{in}(t)$

$$I_p(t) = R|E_{in}(t)|^2, \qquad (4.1)$$

where R is a proportionality constant called photodetector responsivity, measured in units of A/W. In optical systems with intensity modulation and direct detection, opto-electric conversion is carried out by a single photodetector. However, a photodetector is not able to recover the phase of the input field because of the modulus operation in (4.1). Alternatively, coherent receivers resort to much more elaborate structures to recover information conveyed in the amplitude, phase, and polarization of the optical signal.

Coherent receivers are, by definition, structures that mix an input signal at frequency ω_c, with a local oscillator at frequency ω_{LO}, to recover the phase of the input carrier. Mixing is a nonlinear operation that downconverts the signal at frequency ω_c to the intermediate frequency $\omega_{IF} = \omega_c - \omega_{LO}$. Coherent detection can be classified into homodyne, heterodyne, and intradyne. In homodyne detection $\omega_c = \omega_{LO}$, and the signal after mixing is at baseband. In heterodyne detection $\omega_c \neq \omega_{LO}$, and a second downconversion procedure is required to recover the signal from the intermediate frequency ω_{IF} to baseband. Coherent optical systems implement an intradyne detector, for which $\omega_c \approx \omega_{LO}$. In this case, downconversion from $\omega_{IF} \approx 0$ to baseband is carried out by frequency recovery algorithms implemented by DSP.

In optical coherent receivers, the structure that mixes the input optical signal with the local oscillator laser, generating electric currents, is called the optical front-end. A typical optical front-end architecture is shown in Fig. 4.2. The received optical signal $E_r(t)$ is first combined with the local oscillator signal $E_{LO}(t)$ in a 90° hybrid consisting of four 3-dB couplers and one $\pi/2$ phase shifter. To understand the 90° hybrid, let us start with the field transfer function model of a 3-dB coupler [11]

$$\tilde{H}_c = \frac{1}{\sqrt{2}} \begin{bmatrix} 1 & 1 \\ 1 & -1 \end{bmatrix}. \qquad (4.2)$$

Using the transfer in (4.2) it is possible to obtain the electric field expressions of the optical signals at points 1, 2, 3, and 4 indicated in Fig. 4.2, as

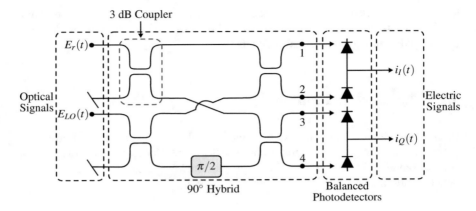

Fig. 4.2 Architecture of a single-polarization optical front-end. The front-end beats the input signal $E_r(t)$ with the local oscillator $E_{LO}(t)$. The 90° hybrid optically combines the input signal and the local oscillator, and the modulus operation of photodetectors provides nonlinearity for frequency mixing. The output of the receiver front-end is currents $i_I(t)$ and $i_Q(t)$ corresponding to the in-phase and quadrature components of the received signal

$$E_1(t) = \frac{1}{2}E_r(t) + \frac{1}{2}E_{LO}(t), \tag{4.3}$$

$$E_2(t) = \frac{1}{2}E_r(t) + \frac{1}{2}E_{LO}(t)e^{j\pi}, \tag{4.4}$$

$$E_3(t) = \frac{1}{2}E_r(t) + \frac{1}{2}E_{LO}(t)e^{j\frac{\pi}{2}}, \tag{4.5}$$

$$E_4(t) = \frac{1}{2}E_r(t) + \frac{1}{2}E_{LO}(t)e^{j\frac{3\pi}{2}}. \tag{4.6}$$

The 90° hybrid combines $E_r(t)$ with $E_{LO}(t)$, producing four outputs mutually phase-shifted by 0, $\pi/2$, π, and $3\pi/2$ rad. The four optical signals are independently received by four photodetectors (also known as two balanced photodetectors [12]), which are combined to produce two electric currents corresponding to in-phase and quadrature components

$$i_I(t) = i_1(t) - i_2(t) = R\,|E_1(t)|^2 - R\,|E_2(t)|^2\,, \tag{4.7}$$

$$i_Q(t) = i_3(t) - i_4(t) = R\,|E_3(t)|^2 - R\,|E_4(t)|^2\,. \tag{4.8}$$

Plugging (4.3)–(4.6) into (4.7) and (4.8) yields

$$i_I(t) = R\left|\frac{1}{2}E_r(t) + \frac{1}{2}E_{LO}(t)\right|^2 - R\left|\frac{1}{2}E_r(t) + \frac{1}{2}E_{LO}(t)e^{j\pi}\right|^2, \tag{4.9}$$

$$i_Q(t) = R \left| \frac{1}{2} E_r(t) + \frac{1}{2} E_{LO}(t) e^{j\frac{\pi}{2}} \right|^2 - R \left| \frac{1}{2} E_r(t) + \frac{1}{2} E_{LO}(t) e^{j\frac{3\pi}{2}} \right|^2. \qquad (4.10)$$

The computation of $i_I(t)$ and $i_Q(t)$ can be simplified by resorting to the following property

$$|z + k|^2 = |z|^2 + |k|^2 + 2\mathbb{R}\{zk^*\}, \qquad (4.11)$$

where z and k are two arbitrary complex numbers, and $\mathbb{R}\{\cdot\}$ takes the real component of $\{\cdot\}$. Finally, by manipulating (4.9) and (4.10), $i_I(t)$ and $i_Q(t)$ can be expressed as

$$i_I(t) = R\mathbb{R}\left\{ E_r(t) E_{LO}^*(t) \right\}, \qquad (4.12)$$

$$i_Q(t) = R\mathbb{R}\left\{ E_r(t) \left[E_{LO}(t) e^{j\pi/2} \right]^* \right\}. \qquad (4.13)$$

So far, we have dealt with the baseband representation of phase- and amplitude-modulated signals, $x_b(t)$, from which the carrier frequency is omitted. If $E_r(t)$ and $E_{LO}(t)$ had the same carrier frequency, the baseband representation of $E_{LO}(t)$ would be a simple constant, and (4.12) and (4.13) would correctly represent the detection process. However, for the modeling of intradyne coherent detection, where $E_r(t)$ and $E_{LO}(t)$ have slightly different carrier frequencies, it is convenient to use the so-called *pre-envelope* or *analytical representation* of $x(t)$, $x_a(t) = x_b(t)e^{i\omega_c t}$, generating, after detection, a signal in intermediate frequency ω_{IF}. While $x(t)$ can be recovered from $x_b(t)$ using $x(t) = \sqrt{2}\mathbb{R}\{x_b(t)e^{i\omega_c t}\}$, it can be recovered from $x_a(t)$ by making $x(t) = \sqrt{2}\mathbb{R}\{x_a(t)\}$. Thus, assuming that $E_r(t)$ has an amplitude modulation $A_r(t)$ and a phase modulation $\phi_r(t)$, it is possible to write the pre-envelope of the received signal as $E_r(t) = A_r(t)e^{j[\phi_r(t)+\omega_c t]}$. In addition, assuming an ideal local oscillator, $E_{LO}(t) = A_{LO}e^{j\omega_{LO}t}$. For the sake of simplicity, we neglect phase and additive noise for now. Equations (4.12) and (4.13) can then be rewritten as

$$i_I(t) = A_r(t)A_{LO}R\mathbb{R}\left\{ e^{j[\phi_r(t)+\omega_{IF}t]} \right\} = A_r(t)A_{LO}R\cos(\phi_r(t) + \omega_{IF}t), \qquad (4.14)$$

$$i_Q(t) = A_r(t)A_{LO}R\mathbb{R}\left\{ e^{j[\phi_r(t)+\omega_{IF}t-\pi/2]} \right\} = A_r(t)A_{LO}R\sin(\phi_r(t) + \omega_{IF}t). \qquad (4.15)$$

Equations (4.14) and (4.15) show that $i_I(t)$ and $i_Q(t)$ recover the real and imaginary parts of $E_r(t)$, and that both terms are multiplied by the responsivity R and the local oscillator amplitude A_{LO}. Indeed, before the invention of the EDFAs, the scaling of detected currents by A_{LO} obtained in coherent detectors was used to improve the receiver sensitivity. In homodyne detection $\omega_{IF} = 0$, and the I and Q

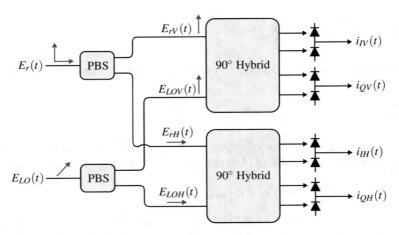

Fig. 4.3 Dual-polarization optical front-end. The received signal $E_r(t)$ and local oscillator $E_{LO}(t)$ are split into V and H polarization orientations by PBSs. Two 90° hybrids combine $E_r(t)$ and $E_{LO}(t)$ in each polarization orientation. Finally, balanced photodetectors produce four current signals, corresponding to the in-phase I and quadrature Q components of the V and H polarization orientations

components of the optical field are seamlessly recovered by the two electric currents. However, as mentioned before, optical systems use intradyne detection, for which the residual ω_{IF} is dynamically estimated by frequency recovery algorithms.

In coherent optical receivers with polarization multiplexing, the receiving structure is more complex than previously described, as shown in Fig. 4.3. The dual-polarization optical front-end is usually constructed by the combination of two single-polarization receivers. Each of these receivers is designed to detect one of the orthogonal polarization orientations of the optical information signal. At first, both $E_r(t)$ and $E_{LO}(t)$ are divided into two orthogonal polarization orientations, here denoted as vertical (V) and horizontal (H). In the first polarization beam splitter (PBS), $E_r(t)$ is decomposed into its V and H components. In the second PBS, the LO signal, which is linearly polarized at 45° with respect to the PBS reference, is also divided into its V and H components. Each of the 90° hybrids combines the corresponding parallel polarization orientations of $E_r(t)$ and E_{LO}. Finally, photodetectors in balanced configuration operate on the hybrids outputs to generate output currents $i_{IV}(t)$, $i_{QV}(t)$, $i_{IH}(t)$, and $i_{QH}(t)$. The random changes in the state of polarization of the received signal cause the V and H polarization orientations to contain a linear combination of the two original transmitted signals. In current coherent optical receivers, these signals are separated by DSP.

4.2 Deskew

The deskew block compensates for a temporal misalignment between in-phase and quadrature components at the output of ADCs, called skew [13]. Skew compensation must be carried out before the in-phase and quadrature components are combined and processed by linear signal processing algorithms. Skew values are usually factory-measured and, in general, do not require estimation. Figure 4.4 illustrates the deskew process. Figure 4.4a shows the in-phase and quadrature components of a received optical signal $E_r(t)$. Figure 4.4b shows the corresponding digital signals $r_I[n]$ and $r_Q[n]$ after analog-to-digital conversion, and illustrates a temporal misalignment between in-phase and quadrature components. Alignment is recovered in Fig. 4.4c using the deskew algorithm. The main difficulty in implementing the deskew block is that the temporal deviation usually corresponds to a fraction of the time between samples.

Any temporal delay τ can be expressed as [1]

$$\tau = nT_{\mathrm{ADC}} + \mu T_{\mathrm{ADC}}, \tag{4.16}$$

where T_{ADC} is the sampling period, n is an integer number of samples, and μ is a fractional delay ($0 < \mu < 1$). While the delay of an integer number of samples n is simple to implement using shift registers, the fractional delay μ requires an interpolator. The principle of interpolation is to approximate a set of received samples $r[n]$ by a continuous-valued function $r(t)$, and resample $r(t)$ at the correct sampling points.

Several solutions have been proposed to carry out skew compensation in optical systems, e.g., in frequency domain [14], or within adaptive equalization [3]. However, the most common solution is perhaps the one based on finite-impulse response (FIR) filters implementing Lagrange interpolation [13, 15]. Given a set of $N + 1$ received samples $r[0] \dots r[N]$ at instants $t[0] \dots t[N]$, the interpolation polynomial at an arbitrary instant t ($t[0] \leq t \leq t[N]$) is given by

$$r(t) = \sum_{n=0}^{N} L_n(t) r[n], \tag{4.17}$$

where the weighting functions, $L_n(t)$, are calculated as

$$L_n(t) = \prod_{m=0,m\neq n}^{N} \frac{t - t[m]}{t[n] - t[m]}. \tag{4.18}$$

As a simple example, a three-tap interpolating function $r(t)$, yielding quadratic interpolator, is given by Tanimura et al. [13]

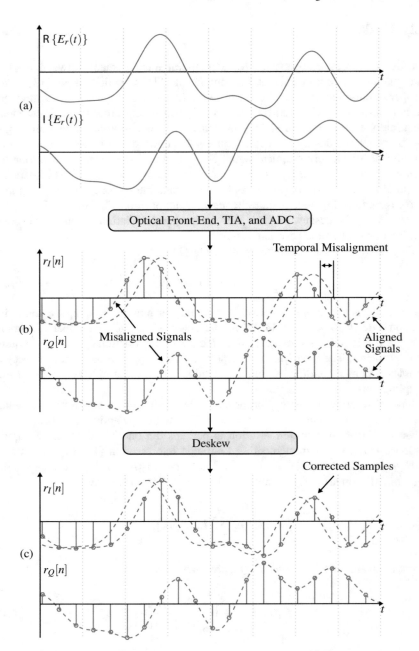

Fig. 4.4 The deskew block. (**a**) In-phase and quadrature components of the received optical signal $E_r(t)$. (**b**) Electric waveforms after analog-to-digital conversion. The sampled waveform $r_I[n]$ becomes temporally displaced with respect to the ideal waveform. (**c**) After deskew, component $r_I[n]$ recovers the alignment with $r_Q[n]$

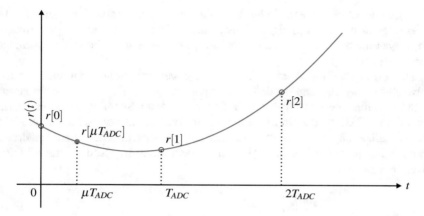

Fig. 4.5 Lagrange interpolator $r(t)$ for quadratic interpolation ($N=2$) of samples $r[0]$, $r[1]$, and $r[2]$, received at instants $t[0] = 0$, $t[1] = T_{ADC}$, and $t[2] = 2T_{ADC}$. The figure also shows value $r(\mu T_{ADC})$, interpolated to correct a fractional time shift of μT_{ADC}

$$r(t) = \left(\frac{t - t[1]}{t[0] - t[1]} \frac{t - t[2]}{t[0] - t[2]} \right) r[0] \tag{4.19}$$

$$+ \left(\frac{t - t[0]}{t[1] - t[0]} \frac{t - t[2]}{t[1] - t[2]} \right) r[1] \tag{4.20}$$

$$+ \left(\frac{t - t[0]}{t[2] - t[0]} \frac{t - t[1]}{t[2] - t[1]} \right) r[2]. \tag{4.21}$$

The interpolation polynomial is constructed in such a way that, for each sample n, the corresponding coefficient is one for $t = t[n]$, and zero for $t = t[m]$, where $m \neq n$. Figure 4.5 shows $r(t)$ for a quadratic interpolation of samples $r[0]$, $r[1]$, and $r[2]$, received at instants $t[0] = 0$, $t[1] = T_{ADC}$, and $t[2] = 2T_{ADC}$. The figure also shows value $r(\mu T_{ADC})$, interpolated to correct a fractional time shift of μT_{ADC}.

Note that (4.17) is seamlessly implemented by a transversal filter structure with tap weights computed using (4.18). Typically, 4- to 10-tap structures are used [16].

4.3 Orthogonalization

Ideally, the in-phase and quadrature components received in the V and H polarization orientations of the receiver front-end are orthogonal, in the sense that they are uncorrelated. However, transmitter or receiver imperfections may cause imbalances that induce a loss of orthogonality [5]. Some examples of imperfections that contribute to these imbalances are the incorrect biasing of in-phase and quadrature modulators, and imperfections in power dividers, optical couplers, and hybrids.

Non-orthogonality between the in-phase and quadrature components translates into distortions in the received signal constellation, affecting the system performance. The higher the order of the modulation format, the more severe are the performance penalties [17].

The orthogonalization stage aims to recover the orthogonal characteristic between the in-phase and quadrature components of the signals. There are several techniques that accomplish this task, being the Gram–Schmidt orthogonalization procedure (GSOP) the most common in optical receivers. In addition to orthogonalization, the GSOP also normalizes the in-phase and quadrature components to unitary power. The set of operations involved in the GSOP is given by Fatadin et al. [5]

$$r_I^{ort}[n] = \frac{r_I[n]}{\sqrt{\mathrm{E}\left\{r_I^2[n]\right\}}}, \tag{4.22}$$

$$r_Q^{int}[n] = r_Q[n] - \frac{\mathrm{E}\left\{r_I[n]r_Q[n]\right\}r_I[n]}{\mathrm{E}\left\{r_I^2[n]\right\}}, \tag{4.23}$$

$$r_Q^{ort}[n] = \frac{r_Q^{int}[n]}{\sqrt{\mathrm{E}\left\{\left(r_Q^{int}[n]\right)^2\right\}}}, \tag{4.24}$$

where $r_I[n]$ and $r_Q[n]$ are the in-phase and quadrature discrete-time sequences obtained after the deskew block, and $E\{\cdot\}$ is the expectation operator.

The GSOP can be interpreted according to the vector representation in Fig. 4.6. First, (4.22) normalizes $r_I[n]$, generating $r_I^{ort}[n]$. Then, (4.23) projects $r_Q[n]$ into the direction orthogonal to $r_I[n]$, generating $r_Q^{int}[n]$. Lastly, $r_Q^{int}[n]$ is also normalized to unitary power, generating $r_Q^{ort}[n]$. Figure 4.7 shows the effect of the GSOP on the signal constellation. In Fig. 4.7a, a QPSK signal contaminated by additive white Gaussian noise has its in-phase and quadrature components correlated because of an imperfect phase shifter at the 90° hybrid. The GSOP is able to restore the original square shape, as shown in Fig. 4.7b.

4.4   Problems

1. A 90° optical hybrid used in a coherent optical front-end has inputs $E_r(t)$ and $E_{LO}(t)$. Using the transfer function of 3-dB couplers, derive the four outputs $E_1(t)$, $E_2(t)$, $E_3(t)$, and $E_4(t)$ of the hybrid.

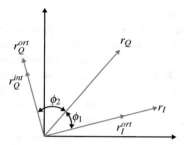

Fig. 4.6 Vectorial representation of the GSOP. The original in-phase and quadrature non-orthogonal components are r_I and r_Q. First, r_I is normalized to unitary power, generating r_I^{ort}. Then, r_Q is projected into the orthogonal direction of r_I, generating r_Q^{int}. Finally, r_Q^{int} is normalized to unitary power, generating r_Q^{ort}

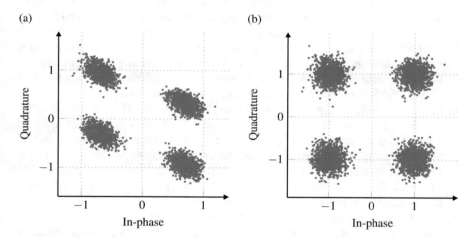

Fig. 4.7 QPSK constellation contaminated by additive white Gaussian noise, distorted because of an imperfect phase shifter at the 90° hybrid. (**a**) Before the GSOP, the in-phase and quadrature components are correlated. (**b**) After the GSOP, the original square constellation in restored

2. Show that balanced photodetectors operating on the outputs of 90° optical hybrids produce two currents corresponding to the in-phase and quadrature components of the received signal $E_r(t)$. Assume that the local oscillator LO operates at the same frequency of the received signal. Use the property that $|z + k|^2 = |z|^2 + |k|^2 + 2\mathbb{R}\{zk^*\}$, where z and k are two arbitrary complex numbers, and $\mathbb{R}\{\cdot\}$ takes the real component of $\{\cdot\}$.

3. Simulate the transmission of Nyquist-shaped QPSK and 16-QAM signals, at the symbol rates of 28 GBaud and 50 GBaud, assuming single-polarization

transmission.[1] Use root-raised cosine (RRC) pulse shape with roll-off factor $\beta^{RC} = 0.1$. Consider OSNRs in the interval from 7 dB to 17 dB. Include in your simulator a MZM, an optical front-end model, and a matched filter at the receiver. Sample the signals at 1 Sa/Symbol and evaluate the BER. Plot simulated curves of BER *versus* OSNR in dB. Compare the results with theoretical curves.

4. Repeat Problem 3 considering an optical transmission system with polarization multiplexing. To do this, generate two signals, one for the V polarization orientation, and the other for the H polarization orientation. At the receiver, insert a dual-polarization optical front-end. Consider OSNRs in the interval from 10 dB to 20 dB.

5. Simulate the transmission of a Nyquist-shaped 16-QAM signal (RRC pulse shape $\beta^{RC} = 0.1$) at the symbol rate of 50 GBaud, assuming single-polarization transmission. Include in your simulator a MZM, an optical front-end model, and a matched filter at the receiver. Insert artificially a skew of 10 ps between the in-phase and quadrature components of the received signal. Simulate the ADC operation by downsampling the received signal at 16 Sa/Symbol to 2 Sa/Symbol. Then, after combining the in-phase and quadrature components, further downsample the signal to 1 Sa/Symbol by selecting the best sampling instant (even or odd samples). Plot the resulting constellation. Simulate the BER *versus* OSNR curve and compare it with the theoretical curve.

6. Implement a Lagrange fourth-order interpolator ($N = 4$) to compensate for the skew generated in the previous problem. The deskew operation should be applied at the signal downsampled to 2 Sa/Symbol. After the deskew operation and combining the in-phase and quadrature components, further downsample the signal to 1 Sa/Symbol by selecting the best sampling instant (even or odd samples). Plot the resulting constellation. Simulate the BER *versus* OSNR curve and compare it with the theoretical curve.

7. Simulate the transmission of a Nyquist-shaped 16-QAM signal (RRC pulse shape with $\beta^{RC} = 0.1$) at the symbol rate of 50 GBaud, assuming a single-polarization orientation. Include in your simulator a MZM modulator, a receiver front-end model, and a matched filter at the receiver. Simulate an imperfection in a phase shifter of the front-end, generating a phase shift of $\phi = (1 + \delta\phi)\pi/2$, where $\delta\phi = 0.1$ is a phase mismatch. Plot the resulting constellation. Simulate the BER *versus* OSNR curve and compare it with the theoretical curve.

8. Apply the GSOP to the constellations obtained in the previous problem. Plot the resulting constellation. Simulate the BER *versus* OSNR curve and compare it with the theoretical curve.

---

[1] Transmit at least 2^{16} symbols per polarization orientation in simulation problems.

4.5 Matlab/Octave Functions

4.5.1 Functions for Sect. 4.1

Matlab/Octave Code 4.1 90° optical hybrid

```
function [E1,E2,E3,E4] = Hybrid90(Er,ELo)
%%%%%%%%%%%%%%%%%%%%%%%%%%%%%%%%%%%%%%%%%%%%%%%%%%%%%%%%%%%%%%%%%%%%%%%%
% HYBRID90 [E1,E2,E3,E4] = Hybrid90(Er,ELo)                          %
%                                                                    %
%   This function simulates a 90 degree hybrid (with ideal components). %
%                                                                    %
% Input:                                                             %
%   Er  = Received optical signal in one pol. orientation (column vector);%
%   ELo = Local oscillator signal in one pol. orientation (column vector);%
%                                                                    %
% Output:                                                            %
%   E1, E2, E3, and E4 = 90 degree hybrid output signals;            %
%                                                                    %
% This function is part of the book Digital Coherent Optical Systems; %
% Darli A. A. Mello and Fabio A. Barbosa;                            %
%%%%%%%%%%%%%%%%%%%%%%%%%%%%%%%%%%%%%%%%%%%%%%%%%%%%%%%%%%%%%%%%%%%%%%%%

    % 3-dB coupler transfer function:
    Hc = (1/sqrt(2))*[1 1; 1 -1];

    % ECouplerTL - Signal at the output of the top-left 3-dB coupler at the
    % 90 degree hybrid;
    ECouplerTL = (Hc*[Er.' ; zeros(1,length(Er))]).';

    % ECouplerBL - Signal at the output of the bottom-left 3-dB coupler at
    % the 90 degree hybrid;
    ECouplerBL = (Hc*[ELo.' ; zeros(1,length(Er))]).';

    % ECouplerTR - Signal at the output of the top-right 3-dB coupler at
    % the 90 degree hybrid;
    ECouplerTR = (Hc*[ECouplerTL(:,1).' ; ECouplerBL(:,1).']).';

    % ECouplerBR - Signal at the output of the bottom-right 3-dB coupler at
    % the 90 degree hybrid;
    ECouplerBR = (Hc*[ECouplerTL(:,2).';ECouplerBL(:,2).'*exp(1i*pi/2)]).';

    % Output signals:
    E1 = ECouplerTR(:,1); E2 = ECouplerTR(:,2); E3 = ECouplerBR(:,1);
    E4 = ECouplerBR(:,2);
end
```

Matlab/Octave Code 4.2 Receiver front-end

```
function [Out] = OpticalFrontEnd(Er,ELo,R,NPol)
%%%%%%%%%%%%%%%%%%%%%%%%%%%%%%%%%%%%%%%%%%%%%%%%%%%%%%%%%%%%%%%%%%%%%%%%
% COHERENTDETECTORFE [Out] = OpticalFrontEnd(Er,ELo,R,NPol)          %
%                                                                    %
%   This function simulates a single- or dual-polarization optical   %
% front-end, considering ideal 90 degree hybrids.                    %
%                                                                    %
% Input:                                                             %
%   Er   = Signal to be detected*;                                   %
%   ELo  = Signal produced by the local oscillator laser;            %
%   R    = Photodetector responsivity;                               %
%   NPol = Number of polarizations;                                  %
%   *Note = For transmission in single pol. orientation 'Er' and 'ELo' must%
```

```
%              be column vectors. For transmission with pol. multiplexing,    %
%              'Er' and 'ELo' must be matrices with with two columns, where    %
%              each column corresponds to the signal of a pol. orientation,    %
%              (V and H pol. orientations);                                    %
%                                                                              %
% Output:                                                                      %
%   Out = Electric currents produced at the front-end. For 'NPol = 1',         %
%         'Out' consists on the electric currents iIV and iQV (column          %
%         vectors). For 'NPol = 2', 'Out' consists on the electric            %
%         currents iIV, iQV, iIH, and iQH (column vectors);                     %
%                                                                              %
% This function is part of the book Digital Coherent Optical Systems;          %
% Darli A. A. Mello and Fabio A. Barbosa;                                      %
%%%%%%%%%%%%%%%%%%%%%%%%%%%%%%%%%%%%%%%%%%%%%%%%%%%%%%%%%%%%%%%%%%%%%%%%%%%%%%%%%%

    % 90 degree Hybrid:
    [E1,E2,E3,E4] = Hybrid90(Er(:,1),ELo(:,1));

    % Photodetection:
    i1 = R*(E1.*conj(E1)) ; i2 = R*(E2.*conj(E2)); % In-phase;
    i3 = R*(E3.*conj(E3)) ; i4 = R*(E4.*conj(E4)); % Quadrature;

    % Balanced photodetection:
    iI = i1 - i2 ; iQ  = i3 - i4;

    % Output signal:
    Out(:,1) = iI ; Out(:,2) = iQ;

    % In the case of pol-mux:
    if NPol == 2
        % 90 degree Hybrid:
        [E1,E2,E3,E4] = Hybrid90(Er(:,2),ELo(:,2));

        % Photodetection:
        i1 = R*(E1.*conj(E1)) ; i2 = R*(E2.*conj(E2)); % In-phase;
        i3 = R*(E3.*conj(E3)) ; i4 = R*(E4.*conj(E4)); % Quadrature;

        % Balanced photodetection:
        iI = i1 - i2 ; iQ  = i3 - i4;

        % Output signal:
        Out(:,3) = iI ; Out(:,4) = iQ;
    end
end
```

Matlab/Octave Code 4.3   Sampling of signals

```
function [r] = ADC(i,SpSIn,NPol,ParamFilter,ParamADC)
%%%%%%%%%%%%%%%%%%%%%%%%%%%%%%%%%%%%%%%%%%%%%%%%%%%%%%%%%%%%%%%%%%%%%%%%%%%%%%%%%%
% ADC [r] = ADC(i,SpSIn,NPol,ParamFilter,ParamADC)                             %
%                                                                              %
%   This function downsamples the signal 'i' so that the output signal 'r'     %
% has 'ParamADC.SpS' samples per symbol. Before downsampling, signal 'i'       %
% is filtered by a filter defined in 'ParamFilter'.                            %
%                                                                              %
% Input:                                                                       %
%   i = Signal to be sampled. For transmission in single pol. orientation,     %
%       'i' must be a matrix with two column vectors. The 1st column must      %
%       have the in-phase component of the signal, while the 2nd column        %
%       must have the quadrature component. For transmission with pol.         %
%       multiplexing, 'i' must be a matrix with four column vectors. The       %
%       1st and 2nd columns must have the in-phase and quadrature              %
%       components of the signal in the V pol. orientation, respectively.      %
%       The 3rd and 4th columns must have the in-phase and quadrature          %
%       components of the signal in the H pol. orientation, respectively;      %
%   SpSIn       = Number of samples per symbol at the input signal 'i';        %
```

```
%    NPol        = Number of polarization orientations used;              %
%    ParamFilter = Struct that specifies parameters of the filter used    %
%                  before sampling 'i', e.g. 'ParamFilter.Type', which can %
%                  be 'RRC','SuperGaussian', or 'NoFiter'.                 %
%      -If ParamFilter.Type = 'RRC', additional required parameters are:  %
%        *ParamFilter.Rolloff: Roll-off of the RRC filter;                %
%        *ParamFilter.Span: Span (in symbols) of the RRC filter           %
%      -If ParamFilter.Type = 'SuperGaussian', additional required        %
%       parameters are:                                                   %
%        *ParamFilter.Order: Order of the SuperGaussian filter            %
%        *ParamFilter.Bw: (Baseband) Filter bandwidth normalized to the   %
%                         symbol rate.                                     %
%      -If ParamFilter.Type = 'NoFilter', the signals are not filtered    %
%         prior to downsampling.                                          %
%    ParamADC    = Struct that specifies parameters of the ADC:           %
%      - ParamADC.SpS = Number of samples per symbol required at the output%
%        signal 'r';                                                      %
%      - ParamADC.FreqError = Deviation in parts per million (ppm) of the %
%        sampling rate. The default is 0 ppm (no sampling errors);        %
%      - ParamADC.PhaseError = Constant lag with the optimal sampling     %
%        instant normalized by the symbol period Ts. Its interval is      %
%        limited to [-Ts/2,Ts/2]. The default value is 0;                 %
%                                                                         %
% Output:                                                                 %
%    r = Signal after filtering and downsampling to 'ParamADC.SpS' samples %
%        per symbol. 'r' is arranged in columns in the same way as 'i';   %
%                                                                         %
% This function is part of the book Digital Coherent Optical Systems;     %
% Darli A. A. Mello and Fabio A. Barbosa;                                 %
%%%%%%%%%%%%%%%%%%%%%%%%%%%%%%%%%%%%%%%%%%%%%%%%%%%%%%%%%%%%%%%%%%%%%%%%%%%%%

    if ~strcmp(ParamFilter.Type,'NoFilter')
        switch ParamFilter.Type
            case 'RRC'
                % Obtaining the filter transfer function:
                g = RRC(ParamFilter.Span,SpSIn,ParamFilter.Rolloff);

                % Filtering the signals:
                i(:,1)=conv(i(:,1),g,'same'); i(:,2)=conv(i(:,2),g,'same');
                if NPol == 2
                i(:,3)=conv(i(:,3),g,'same'); i(:,4)=conv(i(:,4),g,'same');
                end
            case 'SuperGaussian'
                % Frequency vector
                f = fftshift(-0.5:1/length(i(:,1)):0.5-1/length(i(:,1)))...
                    .'*SpSIn;

                % Super-Gaussian filter in the frequency domain:
                G = exp(-log(sqrt(2))*(f/ParamFilter.Bw)...
                    .^(2*ParamFilter.Order));

                % Filtering the signals:
                i(:,1)=ifft(fft(i(:,1)).*G) ; i(:,2)=ifft(fft(i(:,2)).*G);
                if NPol == 2
                    % Filtering the signals:
                    i(:,3)=ifft(fft(i(:,3)).*G); i(:,4)=ifft(fft(i(:,4)).*G);
                end
            otherwise
                error('Filter type not supported');
        end
    end

    % Normalizing the signals to unitary energy:
    i(:,1) = i(:,1)/sqrt(mean(abs(i(:,1)).^2));
    i(:,2) = i(:,2)/sqrt(mean(abs(i(:,2)).^2));
    if NPol == 2
        i(:,3) = i(:,3)/sqrt(mean(abs(i(:,3)).^2));
```

```
            i(:,4) = i(:,4)/sqrt(mean(abs(i(:,4)).^2));
    end

    % Downsampling the signals to 2 samples per symbol:
    PhaseError = 0; % Default value;
    if isfield(ParamADC,'PhaseError')
        PhaseError = ParamADC.PhaseError*SpSIn;
    end
    Len = length(i(:,1)) ; Pos = PhaseError+(1:(SpSIn/ParamADC.SpS):Len)';
    r(:,1) = interp1(1:Len,i(:,1),Pos,'spline','extrap');
    r(:,2) = interp1(1:Len,i(:,2),Pos,'spline','extrap');
    if NPol == 2
        % Downsampling the signals to 2 samples per symbol:
        r(:,3) = interp1(1:Len,i(:,3),Pos,'spline','extrap');
        r(:,4) = interp1(1:Len,i(:,4),Pos,'spline','extrap');
    end

    if isfield(ParamADC,'FreqError')
        Len = length(r(:,1)) ; Pos = (1:1-ParamADC.FreqError:Len)';
        rAux(:,1) = interp1(1:Len,r(:,1),Pos,'spline','extrap');
        rAux(:,2) = interp1(1:Len,r(:,2),Pos,'spline','extrap');
        if NPol == 2
            rAux(:,3) = interp1(1:Len,r(:,3),Pos,'spline','extrap');
            rAux(:,4) = interp1(1:Len,r(:,4),Pos,'spline','extrap');
        end
        r = rAux;
    end
end
```

4.5.2   Functions for Sect. 4.2

Matlab/Octave Code 4.4   Insertion of a temporal misalignment between in-phase and quadrature components of a signal

```
function [Out] = InsertSkew(In,SpS,Rs,NPol,ParamSkew)
%%%%%%%%%%%%%%%%%%%%%%%%%%%%%%%%%%%%%%%%%%%%%%%%%%%%%%%%%%%%%%%%%%%%%%%%%%%%
% INSERTSKEW [Out] = InsertSkew(In,SpS,Rs,NPol,ParamSkew)               %
%                                                                        %
%   This function inserts a temporal misalignment (skew) between in-phase %
% and quadrature components of the signal 'In'. Signal 'In' must be the  %
% signal obtained at the output of the optical front-end, right before   %
% the 'ADC'. In this function, a temporal delay is specified for each    %
% component in 'ParamSkew'. The temporal misalignment between the        %
% components is then applied assuming the minimum temporal delay as      %
% reference. E.g.: A skew of 10 ps between the in-phase and quadrature   %
% components of the signal in V pol. orientation can be specified using  %
% 'ParamSkew.TauIV = 5e-12' and 'ParamSkew.TauQV = -5e-12'. For          %
% transmission with pol. multiplexing, a unique reference is used for the %
% four components (in-phase and quadrature components of each pol.       %
% orientation).                                                          %
%                                                                        %
% Input:                                                                 %
%   In        = Input signals. For transmission in single pol. orientation%
%               'In' must be a matrix with two column vectors. The 1st and%
%               2nd columns must have the in-phase and quadrature        %
%               components of the signal (e.g., at the output of the ADC).%
%               For transmission with pol. multiplexing, 'In' must be a  %
%               matrix with four column vectors. The 1st and the 2nd     %
%               columns must have the in-phase and quadrature components %
%               of the signal received in the V pol. orientation,        %
```

```
%                    respectively. The 3rd and 4th columns have the in-phase   %
%                    and quadrature components of the signal received in the H  %
%                    pol. orientation, respectively;                           %
%    SpS       = Number of samples per symbol in the input signal 'In';        %
%    Rs        = Symbol rate in Symbols/second;                                %
%    NPol      = Number of polarization orientations;                          %
%    ParamSkew = Struct that specifies the temporal delay (in seconds) for     %
%                    each component of the input signal, i.e,                   %
%                        ParamSkew.TauIV and ParamSkew.TauQV;                   %
%                        ParamSkew.TauIH and ParamSkew.TauQH (if NPol == 2);    %
%    *Note: 'In' is the signal obtained at the output of the optical front-     %
%                end, right before the 'ADC'.                                   %
%                                                                              %
% Output:                                                                       %
%    Out = Signal produced after skew insertion. 'Out' is organized in        %
%          columns in the same way as 'In'. E.g., 'Out' is the signal at       %
%          the input of the ADC;                                               %
%                                                                              %
% This function is part of the book Digital Coherent Optical Systems;          %
% Darli A. A. Mello and Fabio A. Barbosa;                                      %
%%%%%%%%%%%%%%%%%%%%%%%%%%%%%%%%%%%%%%%%%%%%%%%%%%%%%%%%%%%%%%%%%%%%%%%%%%%%%%%%%%
    % Input signals:
    iIV = In(:,1) ; iQV = In(:,2);
    if NPol == 2
        iIH = In(:,3) ; iQH = In(:,4);
    end

    % Calculating the interpolation factor given the timing skew
    % (TauIV/H and TauQV/H, for in-phase and quadrature components of
    % V pol. and H pol. orientations, respectively):
    Ts = 1/(SpS*Rs) ; Skew = [ParamSkew.TauIV/Ts ParamSkew.TauQV/Ts];
    if NPol == 2
        Skew = [Skew ParamSkew.TauIH/Ts ParamSkew.TauQH/Ts];
    end

    % Using the min skew as reference:
    Skew = Skew-min(Skew);

    % Inserting skew in the samples:
    Len = length(iIV);
    iIV = interp1(0:Len-1,iIV,Skew(1):Len-1,'spline','extrap').';
    iQV = interp1(0:Len-1,iQV,Skew(2):Len-1,'spline','extrap').';
    if NPol == 2
        iIH = interp1(0:Len-1,iIH,Skew(3):Len-1,'spline','extrap').';
        iQH = interp1(0:Len-1,iQH,Skew(4):Len-1,'spline','extrap').';

        % Output signals with the same length:
        MinLength = min([length(iIV) length(iQV) length(iIH) length(iQH)]);
        Out(:,1) = iIV(1:MinLength) ; Out(:,2) = iQV(1:MinLength);
        Out(:,3) = iIH(1:MinLength) ; Out(:,4) = iQH(1:MinLength);
    else
        % Output signals with the same length:
        MinLength = min([length(iIV) length(iQV)]);
        Out(:,1)  = iIV(1:MinLength) ; Out(:,2) = iQV(1:MinLength);
    end
end
```

Matlab/Octave Code 4.5 Compensation of temporal misalignment between signals

```
function [rOut] = Deskew(rIn,SpSRx,Rs,NPol,N,ParamSkew)
%%%%%%%%%%%%%%%%%%%%%%%%%%%%%%%%%%%%%%%%%%%%%%%%%%%%%%%%%%%%%%%%%%%%%%%%%%%%%%%%%%
% DESKEW [rOut] = Deskew(rIn,SpSRx,Rs,NPol,N,ParamSkew)                        %
%                                                                              %
%   This function performs deskew in the signals 'rIn' using a Lagrange        %
% interpolator of order 'N'. The interpolator is implemented by a FIR          %
```

```
% filter of length 'N+1'. The temporal misalignment is compensated taking %
% into account the lowest temporal delay. The temporal delays of each       %
% of the input signal (in-phase and quadrature components) are specified    %
% in 'ParamSkew'.                                                            %
%                                                                            %
% Input:                                                                     %
%   rIn        = Input signal in which deskew will be performed. For         %
%                transmission in single pol. orientation, 'In' must be a     %
%                matrix with two columns. The 1st and 2nd columns must       %
%                have the in-phase and quadrature components of the signal,  %
%                respectively. For transmission with pol. multiplexing,      %
%                'In' must be a matrix with four columns. The 1st and the    %
%                2nd columns must have the in-phase and quadrature           %
%                components of the signal in V pol. orientation, while the   %
%                3rd and 4th columns, the in-phase and quadrature            %
%                components of the signal in H pol. orientation,             %
%                respectively;                                               %
%   SpSRx      = Samples/symbol at the receiver (after the ADC);             %
%   Rs         = Symbol rate in Symbols/second;                              %
%   NPol       = Number of polarization orientations;                       %
%   N          = Order of the Lagrangean interpolation polynomial;           %
%   ParamSkew = Struct that specifies the temporal delay for each            %
%                component of the input signal:                              %
%                       ParamSkew.TauIV and ParamSkew.TauQV;                 %
%                       ParamSkew.TauIH and ParamSkew.TauQH (if NPol == 2)   %
%                                                                            %
% Output:                                                                    %
%   rOut = Signal after deskew. 'rOut' is arranged in columns in the same    %
%          way as 'rIn';                                                     %
%                                                                            %
% This function is part of the book Digital Coherent Optical Systems;        %
% Darli A. A. Mello and Fabio A. Barbosa;                                    %
%%%%%%%%%%%%%%%%%%%%%%%%%%%%%%%%%%%%%%%%%%%%%%%%%%%%%%%%%%%%%%%%%%%%%%%%%%%%%%%%

    % Skew to be compensated using the TADC as reference:
    TADC=1/(SpSRx*Rs) ; Skew = [ParamSkew.TauIV/TADC ParamSkew.TauQV/TADC];
    if NPol == 2
        Skew = [Skew ParamSkew.TauIH/TADC ParamSkew.TauQH/TADC];
    end

    % Using the min skew as reference:
    Skew = Skew - min(Skew);

    % Integer and fractional part of the skew:
    nTADC = floor(Skew) ; muTADC = -(Skew-nTADC);

    % Obtaining the FIR filter and interpolating the signals:
    NTaps = N+1; % Number of filter taps;
    for i = 1:size(rIn,2)
        L = zeros(NTaps,1) ; Aux = 1;

        % Obtaining the Lagrangean interpolator:
        for n = (0:N) - floor(mean(0:N)) + nTADC(i)
            m     = (0:N) - floor(mean(0:N)) + nTADC(i) ; m(m == n) = [];
            L(Aux) = prod((muTADC(i) - m)./(n - m))      ; Aux = Aux + 1;
        end

        % Interpolating the received signal (sIn):
        sAux = flipud(convmtx([zeros(1,floor(NTaps/2)) rIn(:,i).'...
            zeros(1,floor(NTaps/2))],NTaps));
        sAux = sAux(:,NTaps:end-(NTaps)+1) ; rOut(:,i) = (L.'*sAux).';
    end
end
```

Matlab/Octave Code 4.6 Orthogonalization using the GSOP

```
function [rOut] = GSOP(rIn,NPol)
%%%%%%%%%%%%%%%%%%%%%%%%%%%%%%%%%%%%%%%%%%%%%%%%%%%%%%%%%%%%%%%%%%%%%%%%%%%%
% GSOP [rOut] = GSOP(rIn,NPol)                                            %
%                                                                        %
%   This function performs Gram-Schmidt orthogonalization in signal 'rIn'.%
%                                                                        %
% Input:                                                                 %
%   rIn = Input signal in which orthogonalization will be performed.     %
%         For transmission in single pol. orientation, 'In' must be a    %
%         matrix with two columns. The 1st and 2nd columns must have the %
%         in-phase and quadrature components of the signal, respectively. %
%         For transmission with pol. multiplexing, 'In' must be a matrix %
%         with four columns. The 1st and the 2nd columns must have the   %
%         in-phase and quadrature components of the signal in V pol.     %
%         orientation, while the 3rd and 4th columns, the in-phase and   %
%         quadrature components of the signal in H pol. orientation,     %
%         respectively. 'rIn' is, e.g., the signal at the output of the  %
%         'Deskew' stage;                                                %
%   NPol = Number of polarization orientations used;                    %
%                                                                        %
% Output:                                                                %
%   rOut = Signal after Gram-Schmidt orthogonalization. 'rOut' is arranged%
%          in columns in the same way as 'rIn'. Each signal is normalized %
%          to unitary power;                                             %
%                                                                        %
% This function is part of the book Digital Coherent Optical Systems;    %
% Darli A. A. Mello and Fabio A. Barbosa;                                %
%%%%%%%%%%%%%%%%%%%%%%%%%%%%%%%%%%%%%%%%%%%%%%%%%%%%%%%%%%%%%%%%%%%%%%%%%%%%

        % Taking the in-phase component as reference:
        rIOrt = rIn(:,1)/sqrt(mean(rIn(:,1).^2));

        % Orthogonalization:
        rQInt = rIn(:,2)-mean(rIn(:,1).*rIn(:,2))*rIn(:,1)/mean(rIn(:,1).^2);
        rQOrt = rQInt/sqrt(mean(rQInt.^2));

        % Complex output signal:
        rOut = [rIOrt rQOrt];

        if NPol == 2
          % Taking the in-phase component as reference:
          rIOrt = rIn(:,3)/sqrt(mean(rIn(:,3).^2));

          % Orthogonalization:
          rQInt = rIn(:,4)-mean(rIn(:,3).*rIn(:,4))*rIn(:,3)/mean(rIn(:,3).^2);
          rQOrt = rQInt/sqrt(mean(rQInt.^2));

          % Complex output signal:
          rOut = [rOut rIOrt rQOrt];
        end
end
```

References

1. S.J. Savory, Digital coherent optical receivers: algorithms and subsystems. IEEE J. Sel. Top. Quantum Electron. **16**(5), 1164–1179 (2010)
2. V. Rozental, Hitless rate and bandwidth switching in dynamically reconfigurable coherent optical systems, Ph.D. dissertation (University of Brasilia, Brasilia, 2016). https://repositorio. unb.br/handle/10482/21966

3. R. Rios-Müller, J. Renaudier, G. Charlet, Blind receiver skew compensation and estimation for long-haul non-dispersion managed systems using adaptive equalizer. J. Lightwave Technol. **33**(7), 1315–1318 (2015)
4. E.P. da Silva, D. Zibar, Widely linear equalization for IQ imbalance and skew compensation in optical coherent receivers. J. Lightwave Technol. **34**(15), 3577–3586 (2016)
5. I. Fatadin, S.J. Savory, D. Ives, Compensation of quadrature imbalance in an optical QPSK coherent receiver. IEEE Photon. Technol. Lett. **20**(20), 1733–1735 (2008)
6. S.J. Savory, G. Gavioli, R.I. Killey, P. Bayvel, Electronic compensation of chromatic dispersion using a digital coherent receiver. Opt. Express **15**(5), 2120–2126 (2007)
7. N. Stojanovic, C. Xie, Y. Zhao, B. Mao, N. Gonzalez, J. Qi, N. Binh, Modified Gardner phase detector for Nyquist coherent optical transmission systems, in *Proceedings of Optical Fiber Communication Conference and National Fiber Optic Engineers Conference (OFC/NFOEC)* (Optical Society of America, Washington, 2013), p. JTh2A.50. http://www.osapublishing.org/abstract.cfm?URI=OFC-2013-JTh2A.50
8. V.N. Rozental, T.F. Portela, D.V. Souto, H.B. Ferreira, D.A.A. Mello, Experimental analysis of singularity-avoidance techniques for CMA equalization in DP-QPSK 112-Gb/s optical systems. Opt. Express **19**(19), 18655–18664 (2011). http://www.opticsexpress.org/abstract.cfm?URI=oe-19-19-18655
9. A. Leven, N. Kaneda, U.V. Koc, Y.K. Chen, Frequency estimation in intradyne reception. IEEE Photon. Technol. Lett. **19**(6), 366–368 (2007)
10. E. Ip, J. Kahn, Compensation of dispersion and nonlinear impairments using digital backpropagation. J. Lightwave Technol. **26**(20), 3416–3425 (2008)
11. K.-P. Ho, *Phase-Modulated Optical Communication Systems* (Springer, Boston, 2005)
12. Y. Painchaud, M. Poulin, M. Morin, M. Têtu, Performance of balanced detection in a coherent receiver. Opt. Express **17**(5), 3659–3672 (2009)
13. T. Tanimura, S. Oda, T. Tanaka, T. Hoshida, Z. Tao, J.C. Rasmussen, A simple digital skew compensator for coherent receiver, in *Proceedings of European Conference on Optical Communication (ECOC)* (2009), pp. 1–2
14. N. Stojanovic, X. Changsong, An efficient method for skew estimation and compensation in coherent receivers. IEEE Photon. Technol. Lett. **28**(4), 489–492 (2016)
15. C. Moler, *Numerical Computing with Matlab* (Mathworks, Natick, 2004). http://www.mathworks.com/moler/chapters.html
16. S.M. Bilal, C.R.S. Fludger, Interpolators for digital coherent receivers, in *Proceedings of Photonic Networks; 19th ITG-Symposium* (2018), pp. 1–3
17. M.S. Faruk, S.J. Savory, Digital signal processing for coherent transceivers employing multilevel formats. J. Lightwave Technol. **35**(5), 1125–1141 (2017)

Chapter 5
Equalization

In the previous chapter we studied the opto-electronic conversion process carried out by a digital coherent receiver, and presented the digital signal processing (DSP) blocks of deskew and orthogonalization. The following blocks in the chain of DSP algorithms are two linear equalizers that compensate for the effects of chromatic dispersion (CD) and polarization mode dispersion (PMD).

Before the rebirth of coherent detection, the CD in optical systems with intensity modulation and direct detection was controlled by inserting optical dispersion compensation modules along the optical link. CD management was a relevant topic, and involved the interdependence of CD with nonlinear effects. In digital coherent optical systems, however, CD compensation is completely left to the digital domain. There are several advantages for this choice, such as reducing insertion losses caused by optical CD compensation modules, and facilitating the system design and management. In digital coherent optical receivers, CD is compensated by static filters applied separately to the signals of the two polarization orientations. As CD is practically time-invariant and polarization-independent, the filters corresponding to the two polarization orientations have the same coefficients.

Unlike the static approach used for CD compensation, PMD compensation requires equalizers that are able to adaptively update their coefficients to track variations in the channel frequency response, caused by mechanical perturbations in the fiber geometry or temperature oscillations. Furthermore, as the polarization channels couple during fiber transmission, architectures with multiple inputs a multiple outputs (MIMO) are required. As a side benefit, adaptive equalization also compensates for residual linear effects, such as the CD that is left over from the static equalizer, or narrowband filtering in wavelength-selective switches.

Before the deployment of coherent receivers, PMD was left uncompensated, and polarization multiplexing was commercially unavailable. Therefore, the ability to digitally compensate for PMD, and separate the signals multiplexed in orthogonal polarization orientations, were a major breakthrough in the technology of optical

© Springer Nature Switzerland AG 2021
D. A. de Arruda Mello, F. A. Barbosa, *Digital Coherent Optical Systems*,
Optical Networks, https://doi.org/10.1007/978-3-030-66541-8_5

communications. This chapter details the equalization algorithms implemented in most practical digital coherent optical receivers.

5.1 Static Equalization

The CD compensation module is the first to be implemented in the receiver after the front-end and the compensation of its imperfections. As CD is well-behaved in time and affect equally both polarization orientations, CD compensation filters are static, i.e., their coefficients do not vary with time. The most relevant issues of CD compensation are related to hardware implementation and its complexity and power consumption.

The first task for implementing CD compensation filters is to dimension its parameters taking into account the optical channel properties. Naturally, the equalizer design is directly related to the channel delay spread. As we have seen in Chap. 3, an approximation for the CD delay spread is given by

$$\Delta T = |D|L\Delta\lambda = 2\pi|\beta_2|L\Delta_f, \tag{5.1}$$

where D is the group velocity dispersion (GVD) parameter, L is the fiber length, $\Delta\lambda$ and Δf are the spectral bandwidths in wavelength and frequency, and β_2 is the second derivative of the propagation constant β with respect to the angular frequency ω. If the received signal is sampled at a sampling rate T_{Sa}, the delay spread in number of samples is given by

$$N_{DS} = \left\lceil \frac{\Delta T}{T_{Sa}} \right\rceil = \left\lceil \frac{2\pi|\beta_2|L\Delta f}{T_{Sa}} \right\rceil, \tag{5.2}$$

where T_{Sa} is the sampling period, and $\lceil x \rceil$ denotes the nearest integer larger than x [1]. Assuming minimum-bandwidth Nyquist pulse shaping ($\Delta f \approx R_s$), and an oversampling rate M/K ($T_{Sa} = K/(MR_s)$), (5.2) becomes

$$N_{DS} \approx \left\lceil 2\pi|\beta_2|LR_s^2(M/K) \right\rceil. \tag{5.3}$$

We expect the minimum CD equalizer length to be proportional to N_{DS}, but the exact value depends on many aspects, particularly pulse shape and anti-aliasing filter [1]. An empirical expression to this minimum length, validated for several pulse shapes, is presented in [2] as

$$N_{CD} \approx \left\lceil 6.67|\beta_2|LR_s^2(M/K) \right\rceil. \tag{5.4}$$

Figure 5.1 shows N_{DS} and N_{CD} as a function of the link length for $D = 17$ ps/nm/km, $R_s = 50$ GBd, and an M/K = 2. The CD equalizer can have hundreds

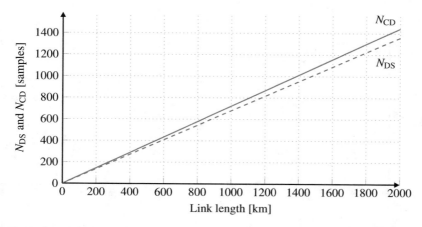

Fig. 5.1 CD delay spread, N_{DS}, and minimum equalizer length, N_{CD}, as a function of the fiber link length, for $D = 17$ ps/nm/km, $R_s = 50$ GBd and an oversampling rate M/K=2

or thousands of taps, depending on the sampling rate and the link length. Therefore, frequency-domain equalization is usually preferred to reduce complexity.

Assuming a Nyquist frequency $f_N = 1/(2T_{Sa})$, the coefficients of a frequency-domain CD equalizer of size N_{FFT} are readily obtained from the CD frequency response

$$H_{CD}[n] = e^{-j\frac{\pi\lambda^2 DL}{c}\left(n\frac{2f_N}{N_{FFT}}\right)^2}, \quad -\frac{N_{FFT}}{2} \leq n \leq \frac{N_{FFT}}{2} - 1. \quad (5.5)$$

Figure 5.2 shows the magnitude and phase responses of a dispersion compensation filter $H_{CD}[n]$, for a link length of $L = 200$ km and $L = 300$ km, dispersion parameter $D = 17$ ps/[nm $\cdot$ km], symbol rate $R_s = 50$ GBd, sampling rate 2 Sa/Symbol, and a filter length of 128 coefficients. As expected, the amplitude diagram is constant, indicating an all-pass filter, while the phase diagram shows a parabolic profile. The highest the accumulated dispersion, the steepest the parabolic profile.

One difficulty in implementing CD compensation in frequency domain is that equalization is performed blockwise. When equalization is carried out in individual blocks, it is not possible to cancel the intersymbol interference (ISI) at the edges of the block. Furthermore, frequency-domain equalization involves calculating the discrete Fourier transform (DFT) of the block, multiplying by the filter coefficients, and calculating the inverse discrete Fourier transform (IDFT), usually implemented by (inverse) fast Fourier transforms (FFTs/IFFTs). This process corresponds to a cyclic convolution of the signal block by the filter coefficients, which causes time aliasing. Recovering the linear convolution of the received sequence by the filter coefficients, from the blockwise cyclic convolution, can be achieved in two ways. The first one is to insert guard intervals or cyclic prefixes between blocks, which wastes transmission bandwidth. The second solution are the overlap-save or overlap-

(a)   (b)

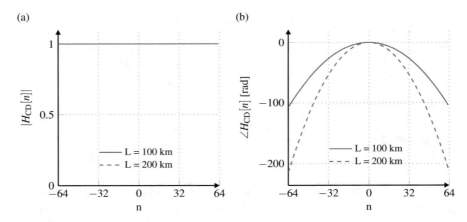

Fig. 5.2 Magnitude and phase of $H_{CD}[n]$, for $D = 17$ ps/nm/km, $L = 100$ km (solid blue line) and $L = 200$ km (dashed orange line), $R_s = 50$ GBd, 2 Sa/Symbol sampling rate and $N_{FFT} = 128$. As expected, the amplitude diagram is constant, indicating an all-pass filter, while the phase diagram shows a parabolic profile

add methods, which do not consume bandwidth but increase DSP complexity. In optical communication systems, overlap-based methods are usually preferred.

The overlap-save method is described in Fig. 5.3 [3]. A sequence of samples enters the receiver continuously. This sequence is split into blocks of equal length for frequency-domain processing, in such a way that block I_n overlaps blocks I_{n-1} and I_{n+1}.[1] After FFT, multiplication by filter coefficients, and IFFT, output block O_n is obtained. If I_n is equalized independently, the ISI generated by neighboring symbols in the edges of O_n is left uncompensated, and the cyclic convolution causes time aliasing. Therefore, the output sequence is recovered by preserving the central portion of O_n and discarding its edges. This approach is repeated sequentially in a way to avoid any loss of information.

Alternatively, the overlap-add method is presented in Fig. 5.4 [3]. In this method, the input sequence is split into non-overlapping blocks. These blocks are then extended by padding zeros at the start and at the end of the sequence, generating overlapping blocks I_n. After FFT, multiplication by filter coefficients, and IFFT, output block O_n is obtained. The original sequence is recovered by summing up the contributions of the overlapped blocks.

The overlap-save and overlap-add methods have two important parameters, the FFT size, N_{FFT}, and the overlap size, $N_{Overlap}$. The minimum $N_{Overlap}$ value is given by the minimum equalizer length (N_{CD}), while N_{FFT} is usually adjusted to minimize complexity. The number of multiplications per symbol required by overlap methods can be computed as [3, 4]

---

[1] Alternative implementations of overlap-based methods consider only one-side overlap with equivalent performance.

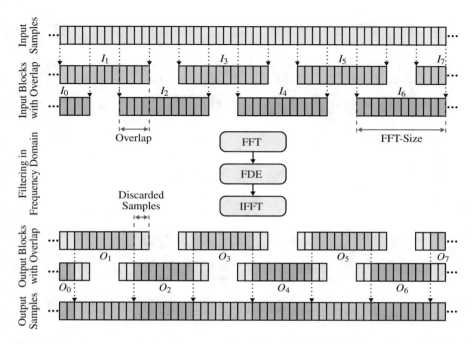

Fig. 5.3 Overlap-save method. A sequence of samples enters the receiver continuously. This sequence is split into blocks of equal length, such that block I_n overlaps blocks I_{n-1} and I_{n+1}. After FFT, multiplication by filter coefficients, and IFFT, output block O_n is obtained. Therefore, output sequence is recovered by preserving the central portion of O_n and discarding its edges [3]. Alternative implementations of overlap-based methods consider only one-side overlap with equivalent performance

$$N_{\text{mult}} = \frac{N_{\text{FFT}}\left[6 \cdot C \cdot \log_2(N_{\text{FFT}}) + 3\right]}{N_{\text{FFT}} - N_{\text{Overlap}} + 1}, \tag{5.6}$$

where C is a constant related to the FFT implementation. For radix-2 implementations (FFT size equal to a power of 2) $C = 1/2$, and for radix-4 implementations (FFT size equal to a power of 4) $C = 3/8$. Figure 5.5 shows the number of multiplications per symbol for standard single-mode fiber (SSMF) propagation ($D = 17$ ps/[nm · km]), $R_s = 50$ GBd and radix-2 FFT implementation. The figure shows that, interestingly, there is an optimal value of N_{FFT} that minimizes complexity, and this value depends strongly on the amount of dispersion to be compensated.

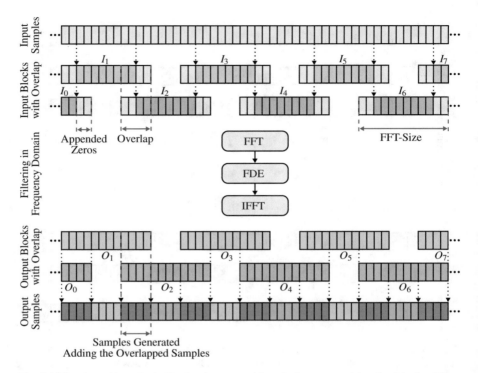

Fig. 5.4 Overlap-add method. The input sequence is split into non-overlapping blocks. These blocks are then extended by padding zeros at the start and at the end of the sequence, generating overlapped blocks I_n. After FFT, multiplication by filter coefficients, and IFFT, output block O_n is obtained. The original sequence is recovered by summing up the contributions of the overlapped blocks [3]. Alternative implementations of overlap-based methods consider only one-side overlap with equivalent performance

5.2   Adaptive Equalization

5.2.1   Fundamentals of Adaptive Equalization

Before delving into the peculiarities of adaptive equalization algorithms used in digital coherent optical receivers, let us first start with some basic principles. We start with the canonical discrete-time channel model shown in Fig. 5.6a [5]. A transmitted signal $s[k]$ is subject to a linear system with impulse response $h[k]$, producing $r[k]$. After that, $r[k]$ is further contaminated by additive white Gaussian noise (AWGN) $\eta[k]$, generating the received signal $x[k]$

$$x[k] = \sum_{i=-\infty}^{\infty} (h[i]\,s[k-i]) + \eta[k] = h[k] * s[k] + \eta[k], \qquad (5.7)$$

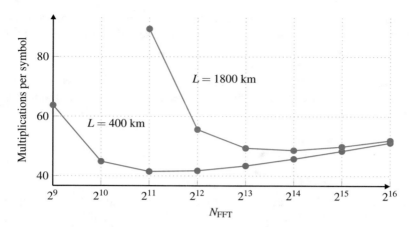

Fig. 5.5 Computational complexity in terms of multiplications per symbol *versus* FFT size, N_{FFT}, for $D = 17$ ps/nm/km, $R_s = 50$ GBd. The overlap size N_{Overlap} was calculated as the delay spread N_{CD} in (5.2). Radix-2 FFT implementation was assumed. The figure indicates the existence of an optimum N_{FFT} that minimizes complexity

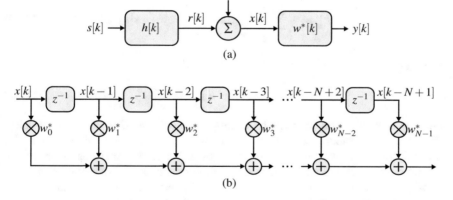

Fig. 5.6 (a) Discrete-time channel model. The transmitted signal $s[k]$ is subject to a linear channel with impulse response $h[k]$, producing $r[k]$. After that, $r[k]$ is further contaminated by AWGN $\eta[k]$, generating the received signal $x[k]$. At the receiver, $x[k]$ is processed by an equalizer with impulse response $w^*[k]$. (b) Transverse linear finite-impulse response filter with input $x[k]$ and coefficients w_i

where ($*$) represents the convolution operation. For a given channel delay d, (5.7) can be rewritten as

$$x[k] = h[d]\,s[k-d] + \sum_{i=-\infty,\,i\neq d}^{\infty} (h[i]\,s[k-i]) + \eta[k]\ . \tag{5.8}$$

Equation (5.8) shows that the received signal $x[k]$ can be expressed as the transmitted signal $s[k]$, delayed by d samples and scaled by $h[d]$, added to noise term $\eta[k]$ and ISI term $\sum\limits_{i=-\infty,\, i\neq d}^{\infty} (h[i]\, s[k-i])$ [6].

At the receiver, we assume that $x[k]$ is processed by a linear transverse filter with finite-impulse response $w^*[k]$ of length N, as depicted in Fig. 5.6b, producing the equalized signal $y[k]$

$$y[k] \;=\; \sum_{i=0}^{N-1} \left(w^*[i]\, x[k-i] \right) = \mathbf{w}^H \mathbf{x}[k], \tag{5.9}$$

where N is the equalizer length, $\mathbf{w} = [w_0 \; w_1 \ldots w_{N-1}]^T$ are the filter coefficients, $\mathbf{x}[k] = [x[k] \; x[k-1] \ldots x[k-N+1]]^T$ is the received signal vector, $(.)^*$ is the complex conjugate operator, and $(.)^H$ is the Hermitian transpose operator. A first attempt to mitigate ISI can be carried out by the so-called zero-forcing (ZF) equalizer. Assuming noiseless transmission, i.e., $\eta = 0$, the equalized signal $y[k]$ can be given in terms of the combined response $c[k]$ of channel and equalizer ($c[k] = h[k] * w^*[k]$)

$$y[k] \;=\; \sum_{i=-\infty}^{\infty} c[i]\, s[k-i] = c[k] * s[k] \;. \tag{5.10}$$

We can define the criterion of perfect equalization as

$$y[k] = \alpha\, s[k-d]\,, \quad \alpha \in \mathbb{R}, \tag{5.11}$$

where α is a gain term (attenuation or amplification), and d is an arbitrary delay. The signal is considered equalized if it recovers the transmitted signal, even if it is multiplied by a constant or delayed in time. Therefore, there are in principle infinite possible equalization solutions, each corresponding to a certain delay and amplification term. Accordingly, aiming at perfect equalization, the equalizer $w[k]$ attempts to approach as close as possible a combined response

$$c[k] = \alpha\, \delta[k-d]\,, \tag{5.12}$$

where $\delta[k]$ is the Kronecker delta. This is called a ZF equalizer because it forces to zero the coefficients of $c[k]$ that do not correspond to delay d. However, ZF equalization is seldom implemented in communications systems because of two main reasons [7]. First, it neglects the presence of noise and, consequently, amplifies the noise components that appear in the vicinity of zeroes in the channel frequency response. Second, it requires the use of an infinite-impulse response (IIR) filter for complete ISI mitigation in finite-impulse response (FIR) channels.

An alternative to the ZF equalizer, which is more suitable for frequency-selective channels with noise, minimizes the mean squared error (MSE) cost function

$$J_{\text{MSE}} = E\left\{ |s[k-d] - y[k]|^2 \right\} = E\left\{ |e[k]|^2 \right\}, \qquad (5.13)$$

where d is an arbitrary delay. Equation (5.13) states that any solution that recovers the original signal $s[k-d]$ is valid, even if it is delayed by d samples. The choice of d can affect the performance of practical (finite length) equalizers. As we will see later, optical systems choose d so as to have the equalizer coefficients reasonably balanced around the center. Plugging (5.9) into (5.13), and choosing $d = 0$ for the sake of clarity, yields

$$
\begin{aligned}
J_{\text{MSE}} &= E\left\{ \left(s[k] - \mathbf{w}^H \mathbf{x}[k] \right) \left(s[k] - \mathbf{w}^H \mathbf{x}[k] \right)^* \right\} \\
&= E\left\{ |s[k]|^2 - s[k]\mathbf{x}[k]^H \mathbf{w} - \mathbf{w}^H \mathbf{x}[k]s^*[k] + \mathbf{w}^H \mathbf{x}[k]\mathbf{x}^H[k]\mathbf{w} \right\} \\
&= E\left\{ |s[k]|^2 \right\} - E\left\{ s[k]\mathbf{x}[k]^H \right\} \mathbf{w} - \mathbf{w}^H E\left\{ \mathbf{x}[k]s^*[k] \right\} \qquad (5.14)\\
&\quad + \mathbf{w}^H E\left\{ \mathbf{x}[k]\mathbf{x}^H[k] \right\} \mathbf{w}.
\end{aligned}
$$

We call $\sigma_s^2 = E\left\{ |s[k]|^2 \right\}$ the power of $s[k]$, $\mathbf{p} = E\{\mathbf{x}[k]s^*[k]\}$ the cross correlation vector of $s[k]$ and $\mathbf{x}[k]$, and $\mathbf{R_x} = E\left\{ \mathbf{x}[k]\mathbf{x}^H[k] \right\}$ the autocorrelation matrix of $\mathbf{x}[k]$. Therefore, (5.14) can be rewritten as

$$J_{\text{MSE}} = \sigma_s^2 - \mathbf{p}^H \mathbf{w} - \mathbf{w}^H \mathbf{p} + \mathbf{w}^H \mathbf{R_x} \mathbf{w}. \qquad (5.15)$$

The gradient of J_{MSE} is obtained by differentiating (5.15) with respect to the filter coefficients $\mathbf{w}$

$$
\begin{aligned}
\nabla J_{\text{MSE}} &= \frac{\partial}{\partial \mathbf{w}} \left\{ \sigma_s^2 - \mathbf{p}^H \mathbf{w} - \mathbf{w}^H \mathbf{p} + \mathbf{w}^H \mathbf{R_x} \mathbf{w} \right\}, \\
&= -2\mathbf{p} + 2\mathbf{R_x}\mathbf{w}. \qquad (5.16)
\end{aligned}
$$

Therefore, J_{MSE} is minimized for [6]

$$\mathbf{w}_{opt} = \mathbf{R_x}^{-1}\mathbf{p}, \qquad (5.17)$$

which is the well-known Wiener filter solution.

In practical applications, the exact values of $\mathbf{R_x}$ and $\mathbf{p}$ are usually not available. Alternatively, it is common to resort to the gradient descendent algorithm to find the optimum filter weights. The gradient descendent algorithm is an iterative algorithm that estimates the gradient of the error after each iteration. A positive gradient for a filter weight w_i indicates that the error increases with an increase in w_i and, therefore, w_i should be reduced. On the other hand, a negative gradient indicates that the error decreases with an increase in w_i and, in this case, w_i should be increased. Following this reasoning, the gradient descendent algorithm updates the filter weights as

$$\mathbf{w}[k] = \mathbf{w}[k-1] - \mu \nabla J_{\text{MSE}}, \tag{5.18}$$

where μ is the step size. The filter coefficients $\mathbf{w}$ become dependent on k as $\mathbf{w}[k] = [w_0[k] \; w_1[k] \ldots w_{N-1}[k]]^T$. However, just as $\mathbf{R_x}$ and $\mathbf{p}$ are not available for calculating of the Wiener solution, they are also not available for calculating ∇J_{MSE} in the gradient descendent algorithm. Instead, a common approach applies the stochastic gradient descent method, which approximates $\mathbf{R_x}$ and $\mathbf{p}$ by instantaneous values $\mathbf{R_x} \approx \mathbf{x}[k]\mathbf{x}[k]^H$, and $\mathbf{p} \approx \mathbf{x}[k]s^*[k]$

$$\nabla J_{\text{MSE}}[k] = 2\mathbf{x}[k]\mathbf{x}^H[k]\mathbf{w}[k-1] - 2\mathbf{x}[x]s^*[k]$$
$$= -2\mathbf{x}[k]e^*[k], \tag{5.19}$$

where $e[k] = s[k] - \mathbf{w}^H[k-1]\mathbf{x}[k] = s[k] - y[k]$. Using the derived expression for $\nabla J_{\text{MSE}}[k]$, the stochastic gradient descendent algorithm updates the filter coefficients as

$$\mathbf{w}[k] = \mathbf{w}[k-1] + \mu\mathbf{x}[k]e^*[k]. \tag{5.20}$$

This is the well-known least mean squares (LMS) algorithm [8]. The step size μ is usually empirically selected to balance excess error and convergence speed. A large μ increases the convergence speed, but can lead to large excess errors (between transmitted and equalized signals) or convergence issues. On the other hand, a small μ reduces the convergence speed and can lead to poor channel tracking in time-varying environments. The LMS algorithm is the basis for most polarization tracking and signal demultiplexing algorithms used in digital coherent optical systems.

5.2.2   Adaptive Equalization in Polarization-Multiplexed Systems

In digital coherent optical receivers, adaptive equalization is usually performed by finite-impulse response filters applied to signals multiplexed in orthogonal polarizations. Therefore, the equalizer has a 2×2 MIMO architecture [9] that is realized using the butterfly structure shown in Fig. 5.7.

The butterfly equalizer is usually implemented by fractional $T/2$-spaced filters to simplify the equalization and synchronization tasks. In this book we use the time variable n for $T/2$-spaced samples, and reserve k for T-spaced samples. The com-

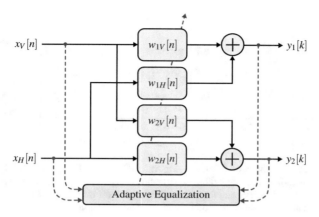

Fig. 5.7 MIMO butterfly equalizer. The complex input signals $x_V[n]$ and $x_H[n]$ are collected at the V and H polarization orientations of the receiver. The butterfly equalizer adapts its coefficients to compensate for polarization effects, producing outputs $y_1[k]$ and $y_2[k]$

plex input signals $x_V[n]$ and $x_H[n]$ are collected at the vertical (V) and horizontal (H) polarization orientations of the receiver. They are linear combinations of the two transmitted signals $s_1[k]$ and $s_2[k]$ that were launched at arbitrary polarization orientations at the transmitter. The butterfly equalizer adapts its coefficients to compensate for polarization effects, producing outputs $y_1[k]$ and $y_2[k]$ that are rotated versions of $s_1[k]$ and $s_2[k]$. The four filters of the butterfly equalizer are given by

$$\mathbf{w}_{1V}[n] = [w_{1V}[n, 0]\ w_{1V}[n, 1]\ w_{1V}[n, 2] \cdots w_{1V}[n, N-1]]^T, \tag{5.21}$$

$$\mathbf{w}_{2V}[n] = [w_{2V}[n, 0]\ w_{2V}[n, 1]\ w_{2V}[n, 2] \cdots w_{2V}[n, N-1]]^T, \tag{5.22}$$

$$\mathbf{w}_{1H}[n] = [w_{1H}[n, 0]\ w_{1H}[n, 1]\ w_{1H}[n, 2] \cdots w_{1H}[n, N-1]]^T, \tag{5.23}$$

$$\mathbf{w}_{2H}[n] = [w_{2H}[n, 0]\ w_{2H}[n, 1]\ w_{2H}[n, 2] \cdots w_{2H}[n, N-1]]^T, \tag{5.24}$$

where $\mathbf{w}_{1V}[n]$, $\mathbf{w}_{2V}[n]$, $\mathbf{w}_{1H}[n]$ e $\mathbf{w}_{2H}[n]$ are the filter coefficients at instant n. These vectors operate at the $T/2$-spaced input vectors $\mathbf{x}_V[n] = [x_V[n]\ x_V[n-1] \ldots x_V[n-N+1]]^T$ and $\mathbf{x}_H[n] = [x_H[n]\ x_H[n-1] \ldots x_H[n-N+1]]^T$. The T-spaced outputs $y_1[k]$ and $y_2[k]$ are obtained as

$$y_1[k] = \mathbf{w}_{1V}^H[n]\mathbf{x}_V[n] + \mathbf{w}_{1H}^H[n]\mathbf{x}_H[n], \tag{5.25}$$

$$y_2[k] = \mathbf{w}_{2V}^H[n]\mathbf{x}_V[n] + \mathbf{w}_{2H}^H[n]\mathbf{x}_H[n], \tag{5.26}$$

where $n = 2k$, assuming equalizers locked to the even samples. There are several algorithms that can be used to adapt the FIR filter coefficients to track the polarization-dependent time-varying effects. Some of these algorithms are

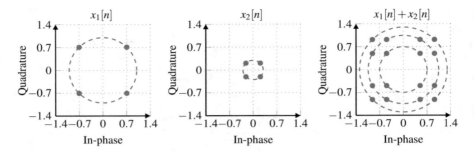

Fig. 5.8 Operating principle of the CMA algorithm. The QPSK signal $x_1[n]$ is added to an interfering signal $x_2[n]$. Although both $x_1[n]$ and $x_2[n]$ have constellations with a constant modulus, the constellation resulting from the sum of both have three radii. Therefore, recovering a signal with a constant modulus is a good indication of equalization

supervised or data-aided, in the sense that they require a training sequence to update the filter coefficients. Others are called unsupervised, non-data aided or blind [10], as they do not require knowledge of the transmitted signal to adapt the filter coefficients. Partially supervised algorithms are common in practice, combining supervised and unsupervised algorithms. The next sections are focused on the most common unsupervised algorithms used in digital coherent optical receivers.

5.2.2.1   Constant Modulus Algorithm (CMA)

In the update equations for the LMS algorithm presented in (5.20), the error signal depends on the knowledge of the transmitted signal $s[k]$. In contrast, unsupervised algorithms apply alternative strategies to provide the error signal. One of these strategies applies a nonlinear function to output y to calculate e and apply the stochastic gradient descent algorithm [11]. Methods that employ this strategy are known as Bussgang algorithms. Among them, the constant modulus algorithm (CMA) is based on the constant modulus criterion.

First proposed by Dominique Godard in 1980 [12], the constant modulus criterion uses a cost function based on high-order statistics of the modulus of y. Its rationale is the fact that, for a constant modulus signal, any type of ISI causes time-varying changes on the modulus and, therefore, avoiding these changes also equalizes the signal. Figure 5.8 illustrates this principle. Suppose that the quadrature phase shift keying (QPSK) signal $x_1[n]$ is added to an interfering QPSK signal $x_2[n]$. Although both $x_1[n]$ and $x_2[n]$ have constellations with constant modulus, the constellation resulting from the sum of both have three radii. Therefore, recovering a signal with constant modulus is a good indication of equalization. The CMA denomination appeared in 1983 [13], when Treichler and Agee independently introduced Godard's criterion [5]. Although it works best with constant modulus signals, e.g., with phase-shift keying modulation, surprisingly, it also exhibits a reasonable performance with non-constant modulus signals, such as 16-ary quadrature amplitude modulation (16-QAM). The main advantage of CMA is its immunity to phase noise, as the algorithm is based on the modulus of the equalized signal, neglecting its phase. When applied to a dual-polarization signal, the CMA is intended to minimize cost functions $J_{CMA,1}$ and $J_{CMA,2}$, related to the outputs of the butterfly equalizer

$$J_{CMA,1} = E\left\{\left(R_d - |y_1[k]|^2\right)^2\right\}, \tag{5.27}$$

$$J_{CMA,2} = E\left\{\left(R_d - |y_2[k]|^2\right)^2\right\}. \tag{5.28}$$

Term R_d is a dispersion constant related to moments of signal $s[k]$ transmitted in each of the polarization orientations [9]

$$R_d = \frac{E\left\{|s[k]|^4\right\}}{E\left\{|s[k]|^2\right\}}. \tag{5.29}$$

For constant modulus constellations with unitary power, $R_d = 1$. The CMA cost function results in an equivalent error signal for the stochastic gradient descendent algorithm of

$$e_{CMA,1}[k] = \left(R_d - |y_1[k]|^2\right) y_1[k], \tag{5.30}$$

$$e_{CMA,2}[k] = \left(R_d - |y_2[k]|^2\right) y_2[k]. \tag{5.31}$$

Thus, using the stochastic gradient descendent algorithm to update the filter coefficients, with the cost function given by the constant modulus criterion, results in the following adaptation rules

$$\mathbf{w}_{1V}[n+2] = \mathbf{w}_{1V}[n] + \mu \mathbf{x}_V[n]\left(R_d - |y_1[k]|^2\right) y_1^*[k], \tag{5.32}$$

$$\mathbf{w}_{2V}[n+2] = \mathbf{w}_{2V}[n] + \mu \mathbf{x}_V[n]\left(R_d - |y_2[k]|^2\right) y_2^*[k], \tag{5.33}$$

$$\mathbf{w}_{1H}[n+2] = \mathbf{w}_{1H}[n] + \mu \mathbf{x}_H[n]\left(R_d - |y_1[k]|^2\right) y_1^*[k], \tag{5.34}$$

$$\mathbf{w}_{2H}[n+2] = \mathbf{w}_{2H}[n] + \mu \mathbf{x}_H[n]\left(R_d - |y_2[k]|^2\right) y_2^*[k], \tag{5.35}$$

where $n = 2k$, assuming equalizers locked to the even samples.

As outputs y_1 and y_2 are updated independently, the CMA can cause both outputs of the butterfly equalizer to converge to the same input signal or to a time-shifted version of the same signal, an effect known as singularity [14]. To circumvent this problem, several solutions have been presented [15–19]. A simple yet effective solution is proposed in [18]. In a first step, the filter coefficients $\mathbf{w}_{1V}[n]$ e $\mathbf{w}_{1H}[n]$

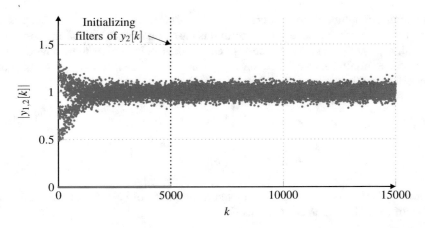

Fig. 5.9 Amplitudes of outputs $y_1[k]$ (blue) and $y_2[k]$ (red) of a MIMO butterfly equalizer operating with QPSK modulation and updated by the CMA. The equalizer starts with the pre-convergence phase of output $y_1[k]$. After 5000 symbols the coefficients $\mathbf{w}_{2V}$ and $\mathbf{w}_{2H}$ are computed by (5.38) and (5.39), and $y_2[k]$ starts to be calculated

are initialized according to the *single spike* configuration, which inserts a single one at the central coefficient of $\mathbf{w}_{1V}[n]$, while all other coefficients are zero. In this way, $\mathbf{w}_{1V}[1]$ and $\mathbf{w}_{1H}[1]$ are given by

$$\mathbf{w}_{1V}[1] = [0 \,\cdots\, 0\;1\;0 \,\cdots\, 0]^T, \tag{5.36}$$

$$\mathbf{w}_{1H}[1] = [0 \,\cdots\, 0\;0\;0 \,\cdots\, 0]^T. \tag{5.37}$$

After this step, the CMA filter update is applied to output $y_1[n]$ until the algorithm converges. After that, the coefficients related to output $y_2[n]$ are initialized as

$$\mathbf{w}_{2H}[n] = \mathbf{w}_{1V}^*[-n], \tag{5.38}$$

$$\mathbf{w}_{2V}[n] = -\mathbf{w}_{1H}^*[-n], \tag{5.39}$$

where $\mathbf{w}[-n]$ is a time-reversed version of filter $\mathbf{w}[n]$. After this initialization process singularities are avoided, and the filters related to both outputs can be updated independently [14]. Figure 5.9 illustrates the convergence phase of a CMA equalizer with singularity avoidance. At the beginning of the convergence stage, only $y_1[n]$ is computed. In the first symbols, at the beginning of the convergence phase, $y_1[n]$ has a wide range of amplitudes. After a few thousand symbols the CMA algorithm converges, and $y_1[n]$ has a well-defined amplitude around one. Finally, after 5000 symbols the coefficients $\mathbf{w}_{2V}$ and $\mathbf{w}_{2H}$ are calculated by (5.38) and (5.39), and $y_2[n]$ starts to be calculated.

5.2.2.2 Radius-Directed Equalization (RDE)

For modulation formats that do not have a constant modulus, the error associated with the CMA cost function is never zero, and the equalization process is sub-optimal [20]. Although M-QAM constellations with M greater than 4 do not have a constant modulus, they can be decomposed into sets of symbols with the same modulus, or radius. For example, Fig. 5.10 indicates the three radii of a 16-QAM constellation. The radius-directed equalization (RDE) algorithm is a modified version of the CMA that takes this feature into account to equalize signals with non-constant modulus [21]. Just like the CMA, the RDE is not affected by phase effects. In RDE, the error signals are given by

$$e_{\mathrm{RDE},1}[k] = \left(R_1[k]^2 - |y_1[k]|^2\right) y_1[k], \tag{5.40}$$

$$e_{\mathrm{RDE},2}[k] = \left(R_2[k]^2 - |y_2[k]|^2\right) y_2[k], \tag{5.41}$$

where $R_1[k]$ and $R_2[k]$ are the symbol radii from the original constellation that are closer to $y_1[k]$ and $y_2[k]$, respectively. Thus, the reference modulus for RDE, $R_1[k]$ and $R_2[k]$, are determined symbol by symbol depending on $y_1[k]$ and $y_2[k]$, unlike CMA, which has a single reference modulus R_d. When applied to the 4-QAM modulation format, the RDE algorithm reduces to the CMA. In the RDE algorithm, the filter coefficients are updated as

$$\mathbf{w}_{1V}[n+2] = \mathbf{w}_{1V}[n] + \mu \mathbf{x}_V[n] \left(R_1[k]^2 - |y_1[k]|^2\right) y_1^*[k], \tag{5.42}$$

$$\mathbf{w}_{2V}[n+2] = \mathbf{w}_{2V}[n] + \mu \mathbf{x}_V[n] \left(R_2[k]^2 - |y_2[k]|^2\right) y_2^*[k], \tag{5.43}$$

$$\mathbf{w}_{1H}[n+2] = \mathbf{w}_{1H}[n] + \mu \mathbf{x}_H[n] \left(R_1[k]^2 - |y_1[k]|^2\right) y_1^*[k], \tag{5.44}$$

$$\mathbf{w}_{2H}[n+2] = \mathbf{w}_{2H}[n] + \mu \mathbf{x}_H[n] \left(R_2[k]^2 - |y_2[k]|^2\right) y_2^*[k]. \tag{5.45}$$

One important limitation of the RDE algorithm is that it does not work properly if the equalized signal is excessively distorted. This happens because the decision process of $R_1[k]$ and $R_2[k]$ is impaired by distortion. Therefore, in severe cases of ISI, the algorithm does not even converge to a valid solution. This problem is usually solved by applying the CMA in a pre-convergence phase, then switching to RDE for continuous operation.

Figure 5.11 shows the evolution of the modulus of outputs $y_1[k]$ (red) and $y_2[k]$ (blue) in the process of convergence. The correct outputs are 16-QAM constellations. In the first stage, only the two filters associated with output $y_1[k]$ are updated using the CMA algorithm. In the second stage, all four filters, associated with outputs $y_1[k]$ and $y_2[k]$, are simultaneously updated. This separation between the first and second stages aims at mitigating singularities, as we explained earlier

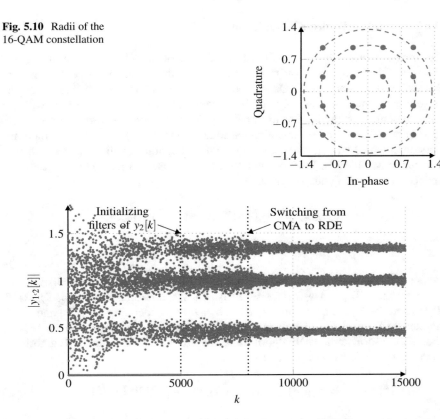

Fig. 5.10 Radii of the 16-QAM constellation

Fig. 5.11 Amplitudes of outputs $y_1[k]$ (blue) and $y_2[k]$ (red) of a MIMO butterfly equalizer operating with 16-QAM modulation. The equalizer starts with the CMA pre-convergence of output $y_1[k]$. The CMA pre-convergence of $y_2[k]$ starts after 5000 symbols to avoid singularities. After 8000 symbols the equalizer switches to the RDE algorithm for continuous operation, reducing the output error

in this section. After this first and second stages, the three radii of the 16-QAM constellation can already be identified, however, there is a large spread around reach radius. The third and last convergence stage switches from CMA to RDE, and the spread around the correct radius is finally reduced, indicating a finer filter convergence.

5.2.2.3   Decision-Directed Algorithm

In the LMS algorithm, the error signal calculation requires knowledge of the transmitted symbol $s[k]$. As $s[k]$ is not available in unsupervised equalization algorithms, one may resort to estimates $\hat{s}[k]$, generated after decision. The decision-directed (DD) equalization algorithm [22] is based on the principle that, under

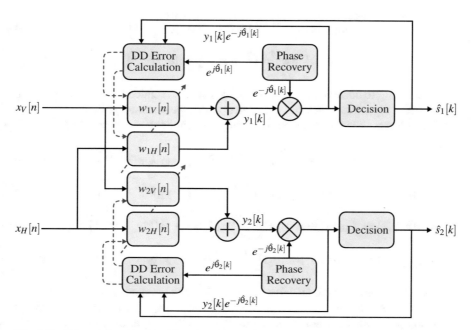

Fig. 5.12 DD equalization algorithm for of a dual-polarization signal. The equalizer outputs $y_1[k]$ and $y_2[k]$ are rotated by phase estimates $-\hat{\theta}_1[k]$ and $-\hat{\theta}_2[k]$, obtained by phase recovery algorithms. The filter coefficients are updated by the stochastic gradient descent algorithm using error signals calculated by decision mismatches and phase noise estimates

normal operation conditions, decisions of equalized symbols are correct with sufficient probability, and may be used as a reference for equalizing future symbols.

Digital coherent optical systems are subject to phase rotations caused by propagation effects, or by imperfections in the transmit and local oscillator lasers. To properly generate estimates $\hat{s}[k]$, these phase rotations must be compensated by phase recovery algorithms. Therefore, the DD algorithm also encompasses a phase recovery algorithm in its feedback loop.

Figure 5.12 shows the DD equalization algorithm for a dual-polarization signal. The equalizer outputs $y_1[k]$ and $y_2[k]$ are rotated by phase estimates $-\hat{\theta}_1[k]$ and $-\hat{\theta}_2[k]$, obtained by phase recovery algorithms (for a detailed discussion on phase recovery algorithms, please refer to Chap. 6). Estimates $\hat{s}_1[k]$ and $\hat{s}_2[k]$ are generated as

$$\hat{s}_1[k] = \left\lfloor y_1[k]e^{-j\hat{\theta}_1[k]} \right\rfloor_D , \tag{5.46}$$

$$\hat{s}_2[k] = \left\lfloor y_2[k]e^{-j\hat{\theta}_2[k]} \right\rfloor_D , \tag{5.47}$$

where $\lfloor \cdot \rfloor_D$ indicates symbol decision. The error signal is then given by Savory et al. [23]

$$e_{\mathrm{DD},1}[k] = e^{j\hat{\theta}_1[k]} \left(\hat{s}_1[k] - y_1[k] e^{-j\hat{\theta}_1[k]} \right), \tag{5.48}$$

$$e_{\mathrm{DD},2}[k] = e^{j\hat{\theta}_2[k]} \left(\hat{s}_2[k] - y_2[k] e^{-j\hat{\theta}_2[k]} \right). \tag{5.49}$$

It is interesting to observe that the error signals are also rotated by the corresponding phase noise signals, as they are calculated after phase recovery. Finally, the filter coefficients are updated using the stochastic gradient descent algorithm as

$$\mathbf{w}_{1V}[n+2] = \mathbf{w}_{1V}[n] + \mu \mathbf{x}_V[n] \left[e^{j\hat{\theta}_1[k]} \left(\hat{s}_1[k] - y_1[k] e^{-j\hat{\theta}_1[k]} \right) \right]^*, \tag{5.50}$$

$$\mathbf{w}_{2V}[n+2] = \mathbf{w}_{2V}[n] + \mu \mathbf{x}_V[n] \left[e^{j\hat{\theta}_2[k]} \left(\hat{s}_2[k] - y_2[k] e^{-j\hat{\theta}_2[k]} \right) \right]^*, \tag{5.51}$$

$$\mathbf{w}_{1H}[n+2] = \mathbf{w}_{1H}[n] + \mu \mathbf{x}_H[n] \left[e^{j\hat{\theta}_1[k]} \left(\hat{s}_1[k] - y_1[k] e^{-j\hat{\theta}_1[k]} \right) \right]^*, \tag{5.52}$$

$$\mathbf{w}_{2H}[n+2] = \mathbf{w}_{2H}[n] + \mu \mathbf{x}_H[n] \left[e^{j\hat{\theta}_2[k]} \left(\hat{s}_2[k] - y_2[k] e^{-j\hat{\theta}_2[k]} \right) \right]^*. \tag{5.53}$$

The convergence of the DD algorithm is related to the quality of estimates $\hat{s}_1[k]$ and $\hat{s}_2[k]$. Therefore it requires a pre-convergence stage, which may be carried out using known symbol sequences or, similarly to the RDE algorithm, using the CMA. The DD equalization algorithm exhibits a suitable performance, even for high-order modulation formats. Its main drawbacks are essentially related to its recursive structure, and the coupling of phase recovery and equalization, which are decoupled in feedforward algorithms such as the CMA or RDE algorithm.

5.3 Problems

1. Plot the curve of delay spread N_{DS} and minimum equalizer length N_{CD} (in number of samples) *versus* symbol rate (20–50 GBd range), for a dispersion-uncompensated 2000-km SSMF link. Assume $D = 17\,\mathrm{ps/nm/km}$, $\lambda = 1550$ nm, minimum-bandwidth Nyquist pulse shaping and an oversampling factor $M/K = 2$.
2. Plot the curve of number of multiplications per symbol N_{mult} *versus* FFT size N_{FFT}, required for CD compensation using overlap-based methods. Assume a dispersion-uncompensated 1000-km SSMF link (D = 17 ps/nm/km), $\lambda = 1550$ nm, minimum-bandwidth Nyquist pulse shaping, oversampling factor $M/K = 2$, a symbol rate of 50 GBd, and Radix-2 FFT implementation. Note that the FFT size should be a power of 2. Calculate the optimum FFT size.

3. Simulate the transmission of a Nyquist-shaped 16-QAM signal at the symbol rate of 50 GBd, assuming single-polarization transmission.[2] Use a root-raised cosine (RRC) pulse shape with roll-off factor $\beta^{RC} = 0.1$ and a span of 20 symbols, assuming 16 Sa/Symbol. Insert the CD corresponding to an uncompensated 1000-km SSMF link ($D = 17\,ps/nm/km$). At the receiver, assume matched filtering. Simulate the ADC operation by downsampling the received signal at 16 Sa/Symbol to 2 Sa/Symbol. Then, after combining the in-phase and quadrature components, further downsample the signal to 1 Sa/Symbol by selecting the best sampling instant (even or odd samples). Plot the BER *versus* OSNR curve.

4. Implement the overlap-save method with optimum FFT size to compensate the CD inserted in the previous item. The overlap-save method should operate at the signal downsampled to 2 Sa/Symbol. Plot the BER *versus* OSNR curve.

5. Simulate the transmission of Nyquist-shaped QPSK signals with polarization multiplexing at the symbol rate of 50 GBaud. For pulse shaping, use the same parameters as in Problem 3. Assume an OSNR of 22 dB. Mix the signals of the two polarization orientations using a rotation angle of 45°. After the optical front-end, use a low-pass tenth order super-Gaussian filter with 25 GHz bandwidth and sample the signal at 2 Sa/Symbol. Separate the polarization-multiplexed signals using the CMA. Implement a 11-tap equalizer and use the single spike initialization. Plot a curve with the time evolution of the modulus of the equalized signal over time, demonstrating the algorithm convergence. Generate four scatter plots, one with the constellation of the first 1000 symbols, and another with the constellation of the last 1000 symbols, for the polarization demultiplexed signals y_1 and y_2.

6. Repeat Problem 5, including uncompensated CD corresponding to a 5-km SSMF with $D = 17\,ps/nm/km$.

7. Repeat Problem 5, now assuming a Nyquist-shaped 16-QAM signals at an OSNR of 28 dB and the separation of the polarization-multiplexed signals using the RDE algorithm. Consider also a pre-convergence stage performed with the CMA.

8. Repeat Problem 7, replacing the mixing of the signals of the two polarization orientations by the insertion of PMD. Simulate the PMD channel of a 300-km SSMF link with 200 sections and average DGD parameter $\overline{|\tau|} = 0.5\,ps/\sqrt{km}$. Plot a curve with the time evolution of the modulus of the equalized signal over time, demonstrating the algorithm convergence. Generate four scatter plots, one with the last 1000 samples of a version downsampled to 1 Sa/Symbol of the signal prior to equalization, and another with the constellation of the last 1000 symbols of the signal after equalization.

---

[2]In all simulation problems, transmit signals with at least 2^{16} symbols per polarization orientation. Consider optical modulation and coherent detection in the simulation models.

5.4   Matlab/Octave Functions

5.4.1   Functions for Sect. 5.1

Matlab/Octave Code 5.1 Compensation of CD using the overlap-save method

```
function [Out] = OverlapSaveCDC(In,D,L,CLambda,Rs,NPol,SpSIn,NFFT,NOverlap)
%%%%%%%%%%%%%%%%%%%%%%%%%%%%%%%%%%%%%%%%%%%%%%%%%%%%%%%%%%%%%%%%%%%%%%%%%%%%%%
% OVERLAPSAVECDC [Out] = OverlapSaveCDC(In,D,L,CLambda,Rs,NPol,SpSIn,...    %
%                        NFFT,NOverlap)                                     %
%                                                                          %
%   This function compensates for chromatic dispersion in the input signal%
% 'In' using the overlap-save method, using FFT of size 'NFFT' and overlap%
% of size 'NOverlap'.                                                      %
%                                                                          %
% Input:                                                                   %
%   In        = Input signal. For transmission in single pol. orientation, %
%               'In' must be a column vector. For transmission with pol.   %
%               multiplexing, 'In' must be a matrix with two column vectors%
%               where each column vector corresponds to the signal of one  %
%               pol. orientation;                                          %
%   D         = Dispersion parameter [ps/(nm*km)];                         %
%   L         = Fiber length [m];                                          %
%   CLambda   = Central wavelength [m];                                     %
%   Rs        = Symbol rate [Sample/symbol];                               %
%   NPol      = Number of polarizations used;                              %
%   SpSIn     = Number of samples per symbol in the input signal 'In';     %
%   NFFT      = FFT size for FDE;                                          %
%   NOverlap  = Overlap size. If 'NOverlap' is odd, it will be forced in   %
%               the code to be the nearest even number greater than the    %
%               configured 'NOverlap';                                     %
%                                                                          %
% Output:                                                                  %
%   Out = Output signal after chromatic dispersion compensation;           %
%                                                                          %
% This function is part of the book Digital Coherent Optical Systems;      %
% Darli A. A. Mello and Fabio A. Barbosa;                                  %
%%%%%%%%%%%%%%%%%%%%%%%%%%%%%%%%%%%%%%%%%%%%%%%%%%%%%%%%%%%%%%%%%%%%%%%%%%%%%%

        % Parameters:
        c = 299792458 ; D = D*1e-6;

        % Index for coefficient calculation and Nyquist frequency:
        n = (-NFFT/2:NFFT/2-1)' ; fN = SpSIn*Rs/2;

        % Calculating the CD frequency response:
        HCD = exp(-1i*pi*CLambda^2*D*L/c*(n*2*fN/NFFT).^2);
        if NPol == 2
            HCD = cat(3,HCD,HCD);
        end

        % Guaranteeing that NOverlap is even:
        NOverlap = NOverlap + mod(NOverlap,2);

        % Extending the input signal so that the blocks can be properly formed:
        AuxLen = size(In,1)/(NFFT-NOverlap);
        if AuxLen ~= ceil(AuxLen)
            NExtra = ceil(AuxLen)*(NFFT-NOverlap)-size(In,1);
            In     = [In(end-NExtra/2+1:end,:); In ; In(1:NExtra/2,:)];
        else
            NExtra = NOverlap;
            In     = [In(end-NExtra/2+1:end,:); In ; In(1:NExtra/2,:)];
        end
```

```matlab
    % Blocks:
    BlocksV = reshape(In(:,1),NFFT-NOverlap,size(In,1)/(NFFT-NOverlap));
    if NPol == 2
        BlocksH = reshape(In(:,2),NFFT-NOverlap,size(In,1)/(NFFT-NOverlap));
        Blocks  = cat(3,BlocksV,BlocksH) ; clearvars BlocksV BlocksH In;
    else
        Blocks = BlocksV ; clearvars BlocksV In;
    end

    % Preallocating the output blocks and the overlap for the first block:
    Out = zeros(size(Blocks)) ; Overlap = zeros(NOverlap,1,NPol);

    % Compensating for the chromatic dispersion:
    for i = 1:size(Blocks,2)
        % Input block with overlap:
        InB = [Overlap ; Blocks(:,i,:)];

        % FFT of the input block:
        InBFreq = fftshift(fft(InB));

        % Filtering in frequency domain:
        OutFDEFreq = InBFreq.*HCD;

        % IFFT of the block after filtering:
        OutFDE = ifft(ifftshift(OutFDEFreq));

        % Overlap:
        Overlap = InB(end-NOverlap+1:end,1,:);

        % Output block:
        OutB = OutFDE(NOverlap/2+1:end-NOverlap/2,1,:);

        % Assigning the samples to the output signal:
        Out(:,i,:) = OutB;
    end

    % Output column vector:
    OutV = reshape(Out(:,:,1),numel(Out(:,:,1)),1);
    if NPol == 2
        OutH = reshape(Out(:,:,2),numel(Out(:,:,2)),1);
        Out  = [OutV OutH];
    else
        Out = OutV;
    end

    % Quantity of samples to discard:
    DInit = 1+(NExtra+NOverlap)/2 ;   DFin = (NExtra-NOverlap)/2;

    % Removing the overlapped zeros:
    Out = Out(DInit:end-DFin,:);
end
```

5.4.2 Functions for Sect. 5.2

Matlab/Octave Code 5.2 Adaptive equalization using the CMA or the RDE algorithms

```
function [y] = AdaptiveEqualizer(x,SpS,ParamDE)
%%%%%%%%%%%%%%%%%%%%%%%%%%%%%%%%%%%%%%%%%%%%%%%%%%%%%%%%%%%%%%%%%%%%%%%%%%%
% ADAPTIVEEQUALIZER [y] = AdaptiveEqualizer(x,SpS,ParamDE)            %
%                                                                    %
%   This function performs equalization in a dual polarization signal 'y' %
% using the CMA or the RDE algorithms. The supported modulation formats %
% are QPSK and 16-QAM. The CMA can be used for pre-convergence of the RDE %
% algorithm.                                                         %
%                                                                    %
% Input:                                                             %
%   x      = Input signal for two pol. orientation (matrix with two  %
%            column vectors, where each column corresponds to the signal %
%            of one pol. orientation). 'x' must be normalized to unitary %
%            power and obtained at 2 Sa/Symbol;                      %
%   SpS    = Number of samples per symbol in the input signal;       %
%  ParamDE =Struct that specifies parameters for the adaptive equalization%
%      - ParamDE.Eq: Defines the algorithm to be used:               %
%                      'CMA' - CMA only;                              %
%                      'RDE' - RDE only;                              %
%                      'CMA+RDE' - CMA is used to initialize the RDE; %
%      - ParamDE.NTaps: Number of taps for the filters in the butterfly %
%                       configuration;                               %
%      - ParamDE.Mu: Step-size for coefficients calculation;         %
%      - ParamDE.SingleSpike: 'true': Single spike initialization;   %
%                             'false': All taps are initialized with zeros%
%      - ParamDE.N1: Number of coefficient calculations to perform prior %
%                    to proper initialization of the filters w2H and w2V; %
%             ('ParamDE.N1' is related to the single spike initialization) %
%      - ParamDE.N2: Number of coefficient calculations to perform prior %
%                    to swicth from CMA to RDE (Note that N2 must only be %
%                    defined if CMA is used for RDE initialization);  %
%      - ParamDE.NOut: Number of samples to discard after equalization; %
%                                                                    %
% Output:                                                            %
%   y = Output signal (at 1 Sa/Symbol) after adaptive equalization;  %
%                                                                    %
% This function is part of the book Digital Coherent Optical Systems; %
% Darli A. A. Mello and Fabio A. Barbosa;                            %
%%%%%%%%%%%%%%%%%%%%%%%%%%%%%%%%%%%%%%%%%%%%%%%%%%%%%%%%%%%%%%%%%%%%%%%%%%%

    % Equalization algorithms:
    CMAFlag = false ; RDEFlag = false ; CMAtoRDE = false ; CMAInit = false;
    if strcmp(ParamDE.Eq,'CMA')
        CMAFlag  = true;
    elseif strcmp(ParamDE.Eq,'RDE')
        RDEFlag  = true;
    elseif strcmp(ParamDE.Eq,'CMA+RDE')
        CMAFlag = true ; CMAtoRDE = true ; CMAInit = true;
    end

    % Radii for constellations with unitary power:
    if CMAFlag
        % Radius for the CMA (E{|x|^4}/E{|x|^2}):
        if ~CMAtoRDE
            R_CMA = 1;
        else
            R_CMA = 1.32;
        end
    end
    if CMAtoRDE || RDEFlag
        R_RDE = [1/sqrt(5) 1 3/sqrt(5)];
    end
```

```
% Important parameters:
Mu    = ParamDE.Mu   ; SingleSpike = ParamDE.SingleSpike;
NTaps = ParamDE.NTaps ; N1 = ParamDE.N1 ; NOut = ParamDE.NOut;
if isfield(ParamDE,'N2') && CMAInit
    N2 = ParamDE.N2;
end

% Input blocks:
x  = [x(end-floor(NTaps/2)+1:end,:) ; x ; x(1:floor(NTaps/2),:)];
xV = convmtx(x(:,1).',NTaps)        ; xH = convmtx(x(:,2).',NTaps);
xV = xV(:,NTaps:SpS:end-NTaps+1) ; xH = xH(:,NTaps:SpS:end-NTaps+1);

% Output length:
OutLength = floor((size(x,1)-NTaps+1)/2) ; clearvars x

% Initializing the outputs
y1 = zeros(OutLength,1) ;  y2 = zeros(OutLength,1);

% Initial filter coefficients:
w1V = zeros(NTaps,1); w1H = zeros(NTaps,1); w2V = zeros(NTaps,1);
w2H = zeros(NTaps,1);

% If single spike initialization:
if SingleSpike
   w1V(floor(NTaps/2)+1) = 1;
end

for i = 1:OutLength
    % Calculating the outputs:
    y1(i) = w1V'*xV(:,i) + w1H'*xH(:,i);
    y2(i) = w2V'*xV(:,i) + w2H'*xH(:,i);

    % Updating the filter coefficients:
    if CMAFlag
        % Constant modulus algorithm:
        [w1V,w1H,w2V,w2H] = CMA(xV(:,i),xH(:,i),y1(i),y2(i),w1V,...
            w1H,w2V,w2H,R_CMA,Mu);

        % Switching from
        if CMAtoRDE
            if i == N2
                CMAFlag = false ; RDEFlag = true;
            end
        end
    elseif RDEFlag
        % Radius-directed equalization:
        [w1V,w1H,w2V,w2H] = RDE(xV(:,i),xH(:,i),y1(i),y2(i),w1V,...
            w1H,w2V,w2H,R_RDE,Mu);
    end

    % Reinitialization of the filter coefficients:
    if i == N1 & SingleSpike
        w2H = conj(w1V(end:-1:1,1)) ; w2V = -conj(w1H(end:-1:1,1));
    end
end

% Output samples:
y = [y1 y2] ; y = y(1+NOut:end,:);
end
```

Matlab/Octave Code 5.3   CMA algorithm

```
function [w1V,w1H,w2V,w2H] = CMA(xV,xH,y1,y2,w1V,w1H,w2V,w2H,R,Mu)
%%%%%%%%%%%%%%%%%%%%%%%%%%%%%%%%%%%%%%%%%%%%%%%%%%%%%%%%%%%%%%%%%%%%%%%%%
% CMA [w1V,w1H,w2V,w2H] = CMA(xV,xH,y1,y2,w1V,w1H,w2V,w2H,R,Mu)        %
%                                                                     %
%  This function performs the update of the filters of the MIMO butterfly %
% equalizer using the CMA algorithm.                                  %
%                                                                     %
% Input:                                                              %
%   xV = Column vector that represents the complex samples at the MIMO %
%        butterfly equalizer input for the vertical pol. orientation; %
%   xH = Column vector that represents the complex samples at the MIMO %
%        butterfly equalizer input for the horizontal pol. orientation; %
%   y1 = Sample at the output 1 of the MIMO butterfly equalizer;      %
%   y2 = Sample at the output 2 of the MIMO butterfly equalizer;      %
%   w1V, w1H, w2V, w2H = N-coefficient FIR filters that compose the MIMO %
%                       butterfly equalizer;                          %
%   R   = radius used as reference for coefficients calculation;      %
%   Mu = Step-size for coefficients calculation;                      %
%   Note: xV and xH must have the same length as the FIR filters w1V, w1H,%
%         w2V, w2H;                                                   %
%                                                                     %
% Output:                                                             %
%   w1V, w1H, w2V, w2H = Updated N-coefficient FIR filters that compose %
%                       the MIMO butterfly equalizer;                 %
%                                                                     %
% This function is part of the book Digital Coherent Optical Systems; %
% Darli A. A. Mello and Fabio A. Barbosa;                             %
%%%%%%%%%%%%%%%%%%%%%%%%%%%%%%%%%%%%%%%%%%%%%%%%%%%%%%%%%%%%%%%%%%%%%%%%%

    %Updating the filters:
    w1V = w1V + Mu*xV*(R-abs(y1).^2)*conj(y1);
    w1H = w1H + Mu*xH*(R-abs(y1).^2)*conj(y1);
    w2V = w2V + Mu*xV*(R-abs(y2).^2)*conj(y2);
    w2H = w2H + Mu*xH*(R-abs(y2).^2)*conj(y2);
end
```

Matlab/Octave Code 5.4   RDE algorithm

```
function [w1V,w1H,w2V,w2H] = RDE(xV,xH,y1,y2,w1V,w1H,w2V,w2H,R,Mu)
%%%%%%%%%%%%%%%%%%%%%%%%%%%%%%%%%%%%%%%%%%%%%%%%%%%%%%%%%%%%%%%%%%%%%%%%%
% RDE [w1V,w1H,w2V,w2H] = RDE(xV,xH,y1,y2,w1V,w1H,w2V,w2H,R,Mu)        %
%                                                                     %
%  This function performs the update of the filters of the MIMO butterfly %
% equalizer using the RDE algorithm.                                  %
%                                                                     %
% Input:                                                              %
%   xV = Column vector that represents the complex samples at the MIMO %
%        butterfly equalizer input for the vertical pol. orientation; %
%   xH = Column vector that represents the complex samples at the MIMO %
%        butterfly equalizer input for the horizontal pol. orientation; %
%   y1 = Sample at the output 1 of the MIMO butterfly equalizer;      %
%   y2 = Sample at the output 2 of the MIMO butterfly equalizer;      %
%   w1V, w1H, w2V, w2H = N-coefficient FIR filters that compose the MIMO %
%                       butterfly equalizer;                          %
%   R   = radius used as reference for coefficients calculation;      %
%   Mu = Step-size for coefficients calculation;                      %
%   Note: xV and xH must have the same length as the FIR filters w1V, w1H,%
%         w2V, w2H;                                                   %
%                                                                     %
% Output:                                                             %
%   w1V, w1H, w2V, w2H = Updated N-coefficient FIR filters that compose %
%                       the MIMO butterfly equalizer;                 %
%                                                                     %
% This function is part of the book Digital Coherent Optical Systems; %
% Darli A. A. Mello and Fabio A. Barbosa;                             %
```

```
%%%%%%%%%%%%%%%%%%%%%%%%%%%%%%%%%%%%%%%%%%%%%%%%%%%%%%%%%%%%%%%%%%%%%%%%%%%%%%
    % Radius for output y1 and output y2:
    [~,r1] = min(abs(R-abs(y1))) ; [~,r2] = min(abs(R-abs(y2)));

    %Updating the filters:
    w1V = w1V + Mu*xV*(R(r1)^2-abs(y1).^2)*conj(y1);
    w1H = w1H + Mu*xH*(R(r1)^2-abs(y1).^2)*conj(y1);
    w2V = w2V + Mu*xV*(R(r2)^2-abs(y2).^2)*conj(y2);
    w2H = w2H + Mu*xH*(R(r2)^2-abs(y2).^2)*conj(y2);
end
```

5.4.3 Auxiliary Functions

Matlab/Octave Code 5.5 Mixing of signals in a polarization-multiplexed transmission using a rotation matrix.

```
function [Out] = RotationMatrix(In,TDegree)
%%%%%%%%%%%%%%%%%%%%%%%%%%%%%%%%%%%%%%%%%%%%%%%%%%%%%%%%%%%%%%%%%%%%%%%%%%%%%%
% ROTATIONMATRIX [Out] = RotationMatrix(In,TDegree)               %
%                                                                 %
%   This function mixes the signals of a polarization-multiplexed %
% transmission applying a rotation matrix                         %
%                   R = [cos(Theta) -sin(Theta)                   %
%                        sin(Theta)  cos(Theta)],                 %
% where Theta is the rotation angle in radians;                   %
%                                                                 %
% Input:                                                          %
%   In      = Input signal. 'In' must be a matrix with two columns, where %
%             each column has the signal of a polarization orientation (V %
%             and H pol. orientations);                           %
%   TDegree = Rotation angle in degrees to be considered in the rotation  %
%             matrix;                                             %
%                                                                 %
% Output:                                                         %
%   Out = Output signals after mixing using the rotation matrix 'R'. 'Out'%
%         is arranged in columns in the same way as 'In';         %
%                                                                 %
% This function is part of the book Digital Coherent Optical Systems; %
% Darli A. A. Mello and Fabio A. Barbosa;                        %
%%%%%%%%%%%%%%%%%%%%%%%%%%%%%%%%%%%%%%%%%%%%%%%%%%%%%%%%%%%%%%%%%%%%%%%%%%%%%%

    % Rotation angle in radians:
    Theta = TDegree*pi/180;

    % Rotation matrix:
    R = [cos(Theta) -sin(Theta); sin(Theta)  cos(Theta)];

    % Applying the rotation to the signals:
    Out(:,1) = R(1,1)*In(:,1) + R(1,2)*In(:,2);
    Out(:,2) = R(2,1)*In(:,1) + R(2,2)*In(:,2);
end
```

References

1. V.N. Rozental, V.E. Parahyba, J.D. Reis, J.F. de Oliveira, D.A.A. Mello, Digital-domain chromatic dispersion compensation for different pulse shapes: practical considerations, in *SBMO/IEEE MTT-S International Microwave and Optoelectronics Conference (IMOC)* (2015), pp. 1–4
2. E. Ip, J. Kahn, Digital equalization of chromatic dispersion and polarization mode dispersion. J. Lightwave Technol. **25**(8), 2033–2043 (2007)
3. T. Xu, G. Jacobsen, S. Popov, M. Forzati, J. Mårtensson, M. Mussolin, J. Li, K. Wang, Y. Zhang, A.T. Friberg, Frequency-domain chromatic dispersion equalization using overlap-add methods in coherent optical system. J. Opt. Commun. **32**(2), 131–135 (2011)
4. T. Xu, G. Jacobsen, S. Popov, J. Li, E. Vanin, K. Wang, A.T. Friberg, Y. Zhang, Chromatic dispersion compensation in coherent transmission system using digital filters. Opt. Express **18**(15), 16243–16257 (2010)
5. V. Rozental, Técnicas para redução de singularidades em receptores ópticos DP-QPSK que utilizam o algoritmo do módulo constante. Master's thesis (University of Brasilia, Brasilia, 2011). https://repositorio.unb.br/handle/10482/9488
6. A. Neves, R. Attux, R. Suyama, M. Miranda, J. Romano, Sobre critérios para equalização não-supervisionada. Controle and Automação **17**(3) (2006)
7. D. Manolakis, V. Ingle, S. Kogon, *Statistical and Adaptive Signal Processing: Spectral Estimation, Signal Modeling, Adaptive Filtering and Array Processing* (Artech House, Norwood, 2005)
8. B. Widrow, M.E. Hoff, Adaptive switching circuits, in *IRE WESCON Convention Record* (1960), pp. 96–104
9. M.S. Faruk, S.J. Savory, Digital signal processing for coherent transceivers employing multilevel formats. J. Lightwave Technol. **35**(5), 1125–1141 (2017)
10. S.J. Savory, Digital coherent optical receivers: algorithms and subsystems. IEEE J. Sel. Top. Quantum Electron. **16**(5), 1164–1179 (2010)
11. S. Haykin, *Unsupervised Adaptive Filtering*, vol. 2 (Wiley, New York, 2000)
12. D. Godard, Self-recovering equalization and carrier tracking in two-dimensional data communication systems. IEEE Trans. Commun. **28**(11), 1867–1875 (1980)
13. J. Treichler, B. Agee, A new approach to multipath correction of constant modulus signals. IEEE Trans. Acoust. Speech Signal Process. **31**(2), 459–472 (1983)
14. V.N. Rozental, T.F. Portela, D.V. Souto, H.B. Ferreira, D.A.A. Mello, Experimental analysis of singularity-avoidance techniques for CMA equalization in DP-QPSK 112-Gb/s optical systems. Opt. Express **19**(19), 18655–18664 (2011). http://www.opticsexpress.org/abstract.cfm?URI=oe-19-19-18655
15. C.B. Papadias, A. Paulraj, A space-time constant modulus algorithm for SDMA systems. Veh. Technol. Conf. **1**(46), 86–90 (1996)
16. A. Vgenis, C.S. Petrou, C.B. Papadias, I. Roudas, L. Raptis, Nonsingular constant modulus equalizer for PDM-QPSK coherent optical receivers. IEEE Photon. Technol. Lett. **22**(1), 45–47 (2010)
17. K. Kikuchi, Polarization-demultiplexing algorithm in the digital coherent receiver, in *IEEE/LEOS Summer Topical Meetings* (2008), pp. 101–102
18. L. Liu, Z. Tao, W. Yan, S. Oda, T. Hoshida, J.C. Rasmussen, Initial tap setup of constant modulus algorithm for polarization de-multiplexing in optical coherent receivers, in *Proceedings of Optical Fiber Communication Conference and National Fiber Optic Engineers Conference (OFC/NFOEC)* (Optical Society of America, New York, 2009), p. OMT2. http://www.osapublishing.org/abstract.cfm?URI=OFC-2009-OMT2
19. C. Xie, S. Chandrasekhar, Two-stage constant modulus algorithm equalizer for singularity free operation and optical performance monitoring in optical coherent receiver, in *Proceedings of Optical Fiber Communication Conference (OFC)* (Optical Society of America, New York, 2010), p. OMK3. http://www.osapublishing.org/abstract.cfm?URI=OFC-2010-OMK3

20. I. Fatadin, D. Ives, S.J. Savory, Blind equalization and carrier phase recovery in a 16-QAM optical coherent system. J. Lightwave Technol. **27**(15), 3042–3049 (2009). http://jlt.osa.org/abstract.cfm?URI=jlt-27-15-3042
21. M.J. Ready, R.P. Gooch, Blind equalization based on radius directed adaptation, in *International Conference on Acoustics, Speech, and Signal Processing*, vol. 3 (IEEE, Albuquerque, 1990), pp. 1699–1702
22. S.U.H. Qureshi, Adaptive equalization, in *Proceedings of the IEEE*, vol. 73(9) (1985), pp. 1349–1387
23. S.J. Savory, G. Gavioli, R.I. Killey, P. Bayvel, Electronic compensation of chromatic dispersion using a digital coherent receiver. Opt. Express **15**(5), 2120–2126 (2007)

Chapter 6
Carrier Recovery

Digital coherent optical systems are subject to phase rotations caused by propagation effects, or by imperfections in the oscillators that generate the carrier signal at the transmitter and the reference (local oscillator) tone at the receiver. Carrier recovery methods estimate these rotations and realign the phase reference of the constellation, such that correct decisions can be made about the transmitted symbols. Before the introduction of digital signal processing (DSP), the traditional method of correcting phase rotations in coherent optical systems was the use of analog phase-locked loops (PLLs) [1]. Analog PLLs directly control the phase of the local oscillator laser to match that of the received carrier but tend to suffer from instabilities and polarization effects, requiring a carefully designed control loop. To circumvent these issues, digital coherent optical systems have free-running local oscillator lasers, and carrier recovery is carried out in the digital domain [2]. There are two main phase effects that affect digital coherent optical systems. The first one is a *carrier frequency offset* [3, 4]. In intradyne coherent detection, transmitter and local oscillator lasers do not operate at exactly the same frequency but exhibit an offset that can reach the order of some few gigahertz. After the optical front-end, this frequency offset is translated to the electric domain, leading to a complete loss of phase reference for the decision process. The second phase effect is that of *phase noise* [5, 6]. In lasers, the spontaneous emission of photons occurring in addition to the required stimulated emissions generates a continuous-time Wiener phase noise process [7]. After the receiver front-end and analog-to-digital conversion, the combined effect of phase noise generated at transmitter and local oscillator lasers produces a discrete-time Wiener phase noise process in digital domain. Phase noise can also arise in long-distance optical systems affected by Kerr nonlinearity [8, 9].

Assuming perfect equalization and polarization demultiplexing, an equalized signal $y[k]$, impaired by a frequency offset, phase noise, and additive noise, can be expressed as

$$y[k] = s[k]e^{j(\theta[k]+k2\pi \Delta f T_s)} + \eta[k], \tag{6.1}$$

© Springer Nature Switzerland AG 2021
D. A. de Arruda Mello, F. A. Barbosa, *Digital Coherent Optical Systems*,
Optical Networks, https://doi.org/10.1007/978-3-030-66541-8_6

where $s[k]$ is the transmitted signal, $\theta[k]$ is the phase noise, Δf is the carrier frequency offset, T_s is the symbol period, and $\eta[k]$ is the additive white Gaussian noise (AWGN). As phase noise and carrier frequency offsets have different properties, they are compensated by different algorithms. *Frequency recovery algorithms* estimate and compensate Δf, while *phase recovery algorithms* estimate and compensate $\theta[k]$. The next sections discuss implementations of these two DSP stages.

6.1 Frequency Recovery

There are several alternatives to implement frequency recovery. It can be implemented in time domain or in frequency domain, and it can be implemented before or after adaptive equalization. Frequency-domain implementations have usually a coarse resolution but can be implemented before adaptive equalization. On the other hand, time-domain implementations have a fine resolution but require the signal to be equalized. A common solution carries out coarse frequency-domain frequency recovery before the adaptive equalizer, and fine time-domain frequency recovery after the adaptive equalizer.

An effective frequency-domain frequency recovery algorithm is based on the spectral analysis of the received signal raised to the 4th power [10]. Although it was originally proposed to be used with the quadrature phase-shift keying (QPSK) modulation format, as described in [11], this algorithm can be extended to M-ary quadrature amplitude modulation (M-QAM) constellations with a tolerable loss in performance. The operating principle of the algorithm can be understood from (6.1). If phase and additive noises are neglected, and $s[k]$ has QPSK modulation, the equalized signal is given by

$$y[k] = e^{j(\frac{\pi}{4} + m[k]\frac{\pi}{2})} e^{jk2\pi \Delta f T_s}, \tag{6.2}$$

where $m[k] = 0, 1, 2, 3$ indicates the quadrant of the QPSK symbol $s[k]$. Raising $y[k]$ to the 4th power yields

$$(y[k])^4 = e^{j\pi} e^{jk2\pi(4\Delta f)T_s}. \tag{6.3}$$

Equation (6.3) indicates that the 4th power operation removes the dependence on $s[k]$ and generates a frequency tone on $4\Delta f$. The frequency offset can then be estimated by calculating this frequency tone and dividing it by four.

The algorithm is presented in Fig. 6.1. First, the input signal is raised to the 4th power. As we have seen, when applied to the QPSK modulation format, this procedure eliminates the signal dependence on the transmitted information. This does not hold for M-QAM modulation with $M > 4$. However, the signal spectrum raised to the 4th power still has a peak at $4\Delta f$, as some partitions of M-QAM constellations are QPSK constellations. As an example, Fig. 6.2 shows the spectrum

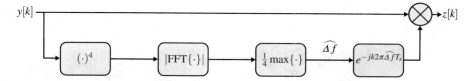

Fig. 6.1 Frequency-domain frequency recovery algorithm. The equalized signal $y[k]$ is first raised to the 4th power to remove the modulation dependency. The estimated frequency offset $\widehat{\Delta f}$ is obtained as 1/4 of the frequency corresponding to the maximum amplitude spectrum. Finally, $\widehat{\Delta f}$ is used to remove the frequency offset

of a 16-QAM signal with a 1-GHz frequency offset on the left, and the spectrum of the same signal, raised to the 4th power, on the right. The spectrum on the right clearly shows a peak around 4 GHz. The estimate of the frequency offset $\widehat{\Delta f}$ is then obtained as

$$\widehat{\Delta f} = \frac{1}{4} \max_f |\text{FFT}\{(y[k])^4\}|, \tag{6.4}$$

where $\text{FFT}\{\cdot\}$ is the fast Fourier transform, and f is the maximal spectral value. Once $\widehat{\Delta f}$ is obtained, the carrier frequency offset can be compensated as

$$z[k] = y[k]e^{-jk2\pi \widehat{\Delta f} T_s}. \tag{6.5}$$

As the same transmitter and local oscillator lasers are used to generate and detect both polarization-multiplexed signals $s_1[k]$ and $s_2[k]$, the frequency offsets experienced by the polarization demultiplexed signals $y_1[k]$ and $y_2[k]$ are similar. Therefore, the estimation of Δf can be performed in only one of the signals ($y_1[k]$ or $y_2[k]$), while the result can used to correct both signals.

One issue that we have not addressed is that the compensation of CD in signals with frequency offset may lead to performance penalties. To avoid such penalties, a coarse frequency recovery stage before CD compensation may be required. A possible solution to this issue is to decouple frequency offset compensation and frequency offset estimation. Frequency offset compensation can be implemented before CD compensation using offset values estimated in previous steps.

6.2 Phase Recovery

Phase noise in optical transmission systems is well-modeled as a discrete-time Wiener stochastic process [12], also known as random walk. In this process, the phase noise $\theta[k]$ is given by

$$\theta[k] = \theta[k-1] + \Delta\theta[k], \tag{6.6}$$

(a)   (b)

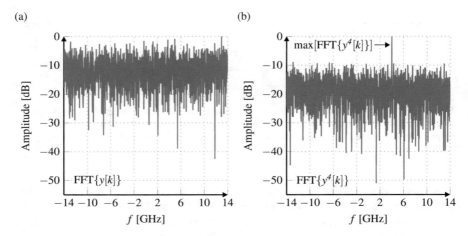

Fig. 6.2 (**a**) Amplitude spectrum of a 16-QAM signal $y[k]$ affected by a frequency offset of $\Delta f = 1$ GHz. (**b**) Amplitude spectrum of $(y[k])^4$. By raising the signal to the 4th power, a frequency tone is observed at $4\Delta f$

where $\Delta\theta[k]$ is a zero-mean Gaussian-distributed random variable. Its variance is given by

$$\sigma_{\Delta\theta}^2 = 2\pi\,\Delta\nu T_s, \tag{6.7}$$

where $\Delta\nu$ is the sum of linewidths of the transmitter and local oscillator lasers. Equation (6.7) indicates that the phase noise intensity is directly proportional to the linewidth of the deployed lasers and inversely proportional to the symbol rate. For this reason, the $\Delta\nu T_s$ product is a common figure of merit to quantify the impact of phase noise on digital coherent optical systems. It is interesting to observe that phase noise is one of the few effects that become less intense with increasing symbol rates. Figure 6.3 shows realizations of the phase noise $\theta[k]$, obtained by two different values of $\Delta\nu T_s$. The higher $\Delta\nu T_s$, the more difficult to recover the carrier.

Phase noise rotates symbols, causing the received constellation to lose the phase reference. Considering the transmission of symbol $s[k]$, the received symbol $z[k]$, contaminated by the phase noise process $\theta[k]$, becomes

$$z[k] = s[k]e^{j\theta[k]} + \eta[k], \tag{6.8}$$

where $\eta[k]$ is the zero-mean AWGN with variance σ_η^2, generated by optical amplification.

Figure 6.4a shows the effect of phase noise with $\Delta\nu T_s = 6.25 \times 10^{-6}$ on 80000 symbols of a 16-QAM constellation. The red crosses represent the original 16-QAM constellation normalized to have unit energy. The blue dots show the constellation rotated by phase noise. Clearly, many of the 80000 symbols already fall outside the

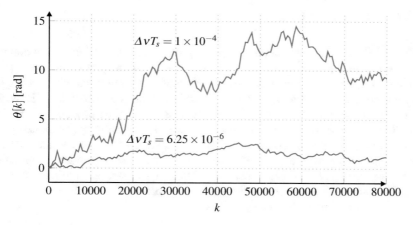

Fig. 6.3 Wiener phase noise realizations for different $\Delta v T_s$ products. A $\Delta v T_s = 1 \times 10^{-4}$ corresponds to 1 MHz lasers ($\Delta v = 2$ MHz) operating on systems at 20 GBaud. A $\Delta v T_s = 6.25 \times 10^{-4}$ corresponds to 100 kHz lasers ($\Delta v = 200$ kHz) operating on systems at 32 GBaud, which is a typical configuration in 100-Gbps transponders

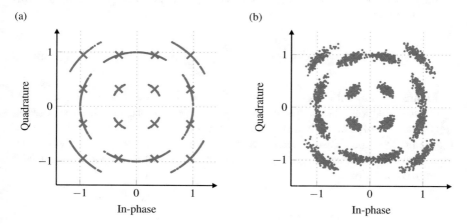

Fig. 6.4 16-QAM constellation with 80000 symbols corrupted by phase noise with $\Delta v T_s = 6.25 \times 10^{-6}$. (**a**) Without additive noise. (**b**) With additive noise. The red crosses indicate the original 16-QAM constellation

correct decision boundaries. Figure 6.4b shows the combined effect of phase noise and AWGN. In this case, in addition to being rotated because of phase noise, the amplitude rings also widen because of the action of additive noise.

6.2.1   The Phase Unwrapper

Before addressing the several phase recovery algorithms applied in optical communications systems, we will discuss a common feature to many of them, which is the phase unwrapper (PU) . We have seen in the previous section that phase noise can be well-modeled by a Wiener stochastic process, which is unconstrained. However, the output of several phase recovery algorithms is a stochastic process limited from $-\pi/P$ to π/P rad, where P is a positive integer. In this case, the estimated phase noise process exhibits discontinuities leading to incorrect phase estimates. These discontinuities are avoided by a PU, an operator that detects sudden jumps in the phase evolution and eliminates them by adding or subtracting multiples of $2\pi/P$ rad. There are several implementations of PUs. A practical solution presented in [2] calculates the unwrapped phase $\hat{\theta}_{\mathrm{PU}}[k]$ as

$$\hat{\theta}_{\mathrm{PU}}[k] = \hat{\theta}[k] + n[k]\frac{2\pi}{P}, \tag{6.9}$$

where $\hat{\theta}[k]$ is the phase noise estimated by the phase recovery algorithm, and $n[k]$ is an integer given by

$$n[k] = \left\lfloor \frac{1}{2} + \frac{\hat{\theta}_{\mathrm{PU}}[k-1] - \hat{\theta}[k]}{2\pi/P} \right\rfloor. \tag{6.10}$$

Figure 6.5 shows the operation of a PU for $P = 4$, which is the value used in M-QAM constellations. The output of the phase recovery algorithm is limited between $+\pi/4$ and $-\pi/4$. The blue dashed curve shows the estimated phase noise process before the PU, and the solid red curve shows the process after phase unwrapping. Phase unwrapping eliminates discontinuities, producing at its output a process with unlimited excursion.

6.2.2   The Viterbi & Viterbi Algorithm

The most simple phase recovery algorithm used in optical communications is the Viterbi & Viterbi algorithm [13]. This technique, developed for M-PSK signals, is based on raising an M-PSK signal to the Mth power to remove data modulation from the received signal. An M-PSK signal $s[k]$ is given by [14]

$$s[k] = \sqrt{E_s}e^{j2m[k]\pi/M+j\pi/M}, \quad m[k] \in 0, 1, \ldots, M-1, \tag{6.11}$$

where E_s is the symbol energy. Raising $s[k]$ to the Mth power yields

$$s^M[k] = E_s^{M/2}e^{j2m[k]\pi+j\pi},$$

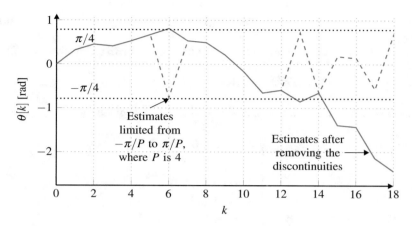

Fig. 6.5 Estimated phase noise before (blue dashed line) and after (solid red line) the PU. The example assumes $P = 4$, which is the value used for square M-QAM constellations

$$s^M[k] = E_s^{M/2} e^{j\pi}.$$ (6.12)

This operation eliminates data modulation of $s[k]$. In the presence of zero-mean AWGN $\eta[k]$ with variance σ_η^2, and multiplicative Wiener phase noise $\theta[k]$, the received signal $z[k]$, raised to the Mth power, becomes

$$z^M[k] = (s[k]e^{j\theta[k]} + \eta[k])^M,$$

$$\approx E_s^{M/2} e^{j\pi} e^{jM\theta[k]} + \zeta[k],$$ (6.13)

where $\zeta[k]$ is a zero-mean AWGN with variance $\sigma_\zeta^2 = M^2 E_s^{M-1} \sigma_\eta^2$ [15, 16]. In the Viterbi & Viterbi algorithm, the estimation of $\theta[k]$ is performed using a vector of symbols to filter out the influence of $\zeta[k]$

$$\mathbf{z} = [z^M[k-N] \cdots z^M[k-1] \, z^M[k] \, z^M[k+1] \cdots z^M[k+N]]^T.$$ (6.14)

Estimating $\theta[k]$ involves, in addition to the current received symbol $z[k]$, N past symbols and N future symbols, comprising a filter of length $L = 2N + 1$. It can be shown that the maximum likelihood (ML) estimate of $\theta[k]$, given $\mathbf{z}$, is calculated as

$$\hat{\theta}_{\text{ML}}[k] = \max_{\theta[k]} \left[f_{\mathbf{z}|\theta[k]}(\mathbf{z}|\theta[k]) \right],$$

$$= \frac{1}{M} \arg\left(\mathbf{w}_{\text{ML}}^T \cdot \mathbf{z} \right) - \frac{\pi}{M},$$ (6.15)

where $f_{\mathbf{z}|\theta[k]}(\mathbf{z}|\theta[k])$ is the probability density function of $\mathbf{z}$ conditioned on $\theta[k]$. The maximum likelihood filter $\mathbf{w}_{\text{ML}}$ is obtained as [16]

$$\mathbf{w}_{ML} = (\mathbf{1}^T \mathbf{C}^{-1})^T, \tag{6.16}$$

where $\mathbf{1}^T$ is an all-ones column vector. The covariance matrix $\mathbf{C}$ is given by [2]

$$\mathbf{C} = E_s^M M^2 \sigma_{\Delta\theta}^2 \mathbf{K}_{L \times L} + E_s^{M-1} M^2 \sigma_\eta^2 \mathbf{I}_{L \times L},$$

where $\mathbf{I}_{L \times L}$ is an identity matrix of size L, and $\mathbf{K}_{L \times L}$ is given by

$$\mathbf{K} = \begin{bmatrix} N & \cdots & 2 & 1 & 0 & 0 & 0 & \cdots & 0 \\ \vdots & \ddots & \vdots & \vdots & \vdots & \vdots & \vdots & \iddots & \vdots \\ 2 & \cdots & 2 & 1 & 0 & 0 & 0 & \cdots & 0 \\ 1 & \cdots & 1 & 1 & 0 & 0 & 0 & \cdots & 0 \\ 0 & \cdots & 0 & 0 & 0 & 0 & 0 & \cdots & 0 \\ 0 & \cdots & 0 & 0 & 1 & 1 & \cdots & 1 \\ 0 & \cdots & 0 & 0 & 1 & 2 & \cdots & 2 \\ \vdots & \iddots & \vdots & \vdots & \vdots & \vdots & \vdots & \ddots & \vdots \\ 0 & \cdots & 0 & 0 & 1 & 2 & \cdots & N \end{bmatrix}.$$

The shape of $\mathbf{w}_{ML}$ depends on the balance of the additive and multiplicative noise variances σ_η^2 and $\sigma_{\Delta\theta}^2$. Figure 6.6 shows the ML filter in different conditions of additive and phase noise. The lower the optical signal-to-noise ratio (OSNR), the flatter the filter, as more symbols are required to filter out the additive noise contribution. On the other hand, the higher the phase noise, the more abrupt the filter edges, as the phase noise samples at the filter edges are less correlated with the central phase noise term.

The Viterbi & Viterbi algorithm is summarized by the block diagram in Fig. 6.7. First, the received signal vector is raised to the Mth power and multiplied by the transpose of the ML filter coefficients. Then, the algorithm calculates the argument of the resulting value, divides it by M, and subtracts π/M. In general, the argument operation yields values that are limited between $-\pi$ and π. Therefore, after division by M, the output is limited between $-\pi/M$ and π/M. Discontinuities are avoided by sending the estimated phases $\hat{\theta}_{ML}[k]$ to a PU that produces outputs $\hat{\theta}_{PU}[k]$ [2]. Finally, the phase noise process is compensated as

$$v[k] = z[k]e^{-j\hat{\theta}_{PU}[k]}. \tag{6.17}$$

6.2.3   Decision-Directed Algorithm

Decision-directed estimators use decisions to eliminate dependence on the transmitted information. In the case of decision-directed (DD) phase recovery, N previous

(a)

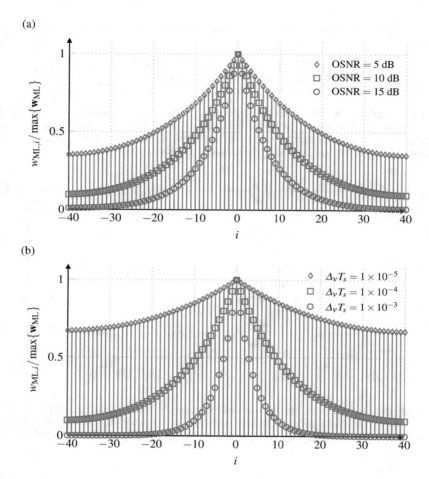

(b)

Fig. 6.6 Coefficients $\mathbf{w}_{\mathrm{ML}} = (\mathbf{1}^T \mathbf{C}^{-1})^T$ of the ML filter used for phase noise estimation in the Viterbi & Viterbi algorithm with $L = 81$. (**a**) With constant $\Delta_v T_s = 1 \times 10^{-4}$ and varying OSNR. (**b**) With constant OSNR = 10 dB and varying $\Delta_v T_s$. We assume a polarization-multiplexed QPSK signal with symbol rate $R_s = 28$ GBd, and symbol energy $E_s = 1$. The lower the OSNR, the flatter the filter, as more symbols are required to filter out the additive noise contribution. On the other hand, the higher the phase noise, the more abrupt the filter edges, as the phase noise samples at the filter edges are less correlated with the central phase noise term

symbol decisions are used to estimate the phase noise realization $\theta[k]$. Estimated phase $\hat{\theta}[k]$ then compensates for the phase shift imprinted on the received symbol $z[k]$. Although fairly simple, this recursive structure has two main drawbacks. First, the algorithm requires several previous decisions to estimate the current phase, which may be a problem for hardware parallelism. Second, the algorithm is based on previous estimates of phase noise (current and future samples are not used), introducing a prediction error that is not present in feedforward solutions. Figure 6.8

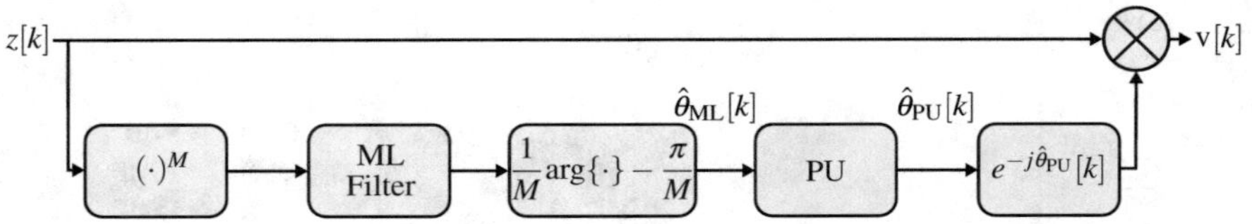

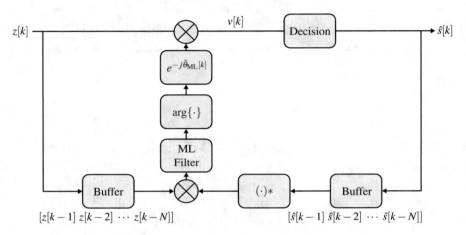

Fig. 6.7 Block diagram of the Viterbi & Viterbi algorithm. First, the received signal vector is raised to the Mth power and multiplied by transpose of the ML filter coefficients. Then, the algorithm calculates the argument of the resulting value, divides it by M, and subtracts π/M. Discontinuities are avoided by sending the estimated phases $\hat{\theta}_{\mathrm{ML}}[k]$ to a PU

Fig. 6.8 DD phase recovery. The algorithm uses a vector of received symbols $\mathbf{z} = [z[k-1]\, z[k-2]\, \cdots\, z[k-N]]^T$ and a vector of decided symbols $\hat{\mathbf{s}} = [\hat{s}[k-1]\, \hat{s}[k-2]\, \cdots\, \hat{s}[k-N]]^T$ to generate estimated phase $\hat{\theta}_{\mathrm{ML}}[k]$. Past phase estimates are weighed by a maximum likelihood filter

shows the architecture of a DD phase recovery solution. The algorithm uses a vector of received symbols $\mathbf{z} = [z[k-1]\, z[k-2]\, \cdots\, z[k-N]]^T$ and a vector of decided symbols $\hat{\mathbf{s}} = [\hat{s}[k-1]\, \hat{s}[k-2]\, \cdots\, \hat{s}[k-N]]^T$ to generate estimated phase $\hat{\theta}_{\mathrm{ML}}[k]$ as

$$\hat{\theta}_{\mathrm{ML}}[k] = arg\left(\sum_{i=1}^{N} w_{\mathrm{ML},i}\, z[k-i]\hat{s}^*[k-i]\right), \tag{6.18}$$

where $w_{\mathrm{ML},i}$ is the ith coefficient of the ML filter. Just as for the Viterbi & Viterbi algorithm, for the DD algorithm the maximum likelihood filter is calculated as $\mathbf{w}_{\mathrm{ML}} = (\mathbf{1}^T \mathbf{C}^{-1})^T$, where $\mathbf{1}^T$ is an all-ones column vector. The covariance matrix $\mathbf{C}$ is given by [15, 16]

$$\mathbf{C} = E_s^2 \sigma_{\Delta\theta}^2 \mathbf{K}_{NxN} + E_s \sigma_\eta^2 \mathbf{I}_{NxN},$$

where $\mathbf{I}_{NxN}$ is an identity matrix of size N, and $\mathbf{K}_{NxN}$ is given by

$$\mathbf{K} = \begin{bmatrix} 0 & 0 & 0 & \cdots & 0 \\ 0 & 1 & 1 & \cdots & 1 \\ 0 & 1 & 2 & \cdots & 2 \\ \vdots & \vdots & \vdots & \ddots & \vdots \\ 0 & 1 & 2 & \cdots & N-1 \end{bmatrix}.$$

The shape of $\mathbf{w}_{ML}$ depends on the balance of the additive and multiplicative noise variances σ_η^2 and $\sigma_{\Delta\theta}^2$. Unlike in the Viterbi & Viterbi algorithm, since only previous symbols are available to estimate $\theta[k]$, $\mathbf{w}_{ML}$ is causal. Figure 6.9 shows the ML filter for the DD algorithm for different conditions of additive and phase noise. Again, the lower the OSNR, the flatter the filter, as more symbols are required to filter out the additive noise contribution. On the other hand, the higher the phase noise, the more abrupt the filter edges, as the phase noise samples are less correlated. As a final step, phase noise is compensated as

$$v[k] = z[k]e^{-j\hat{\theta}_{ML}[k]}. \tag{6.19}$$

6.2.4   The Blind Phase Search Algorithm

High-order M-QAM formats have a reduced tolerance to phase noise due to the small Euclidean distance between their constellation points. DD phase recovery schemes exhibit a reasonable performance for M-QAM constellations, but the feedback loop may pose implementation issues considering the high degree of parallelism required in practical application-specific integrated circuit (ASIC) deployments. Alternatively, the blind phase search (BPS) algorithm [5] achieves excellent performance with high-order M-QAM constellations using a feedforward architecture that is suitable for parallel implementations.

Figure 6.10 shows the BPS algorithm architecture. The first step of the algorithm is to apply B test rotations θ_b to the received signal $z[k]$. The range of rotations p is equal to the angle of ambiguity of the signal constellation. For M-QAM formats, $p = \pi/2$. Then, for an even B, the rotation angles θ_b are calculated as

$$\theta_b = \frac{b}{B}p, \quad b \in \{-\frac{B}{2}, \cdots, -1, 0, 1, \cdots, \frac{B}{2} - 1\}. \tag{6.20}$$

The rotated symbol $z_b[k]$ is given by

$$z_b[k] = z[k]e^{j\theta_b}. \tag{6.21}$$

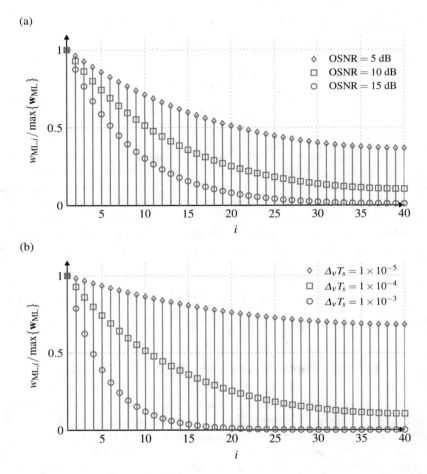

Fig. 6.9 Coefficients $\mathbf{w}_{ML} = (\mathbf{1}^T \mathbf{C}^{-1})^T$ of the ML filter used for phase noise estimation in the DD algorithm with N = 40. (**a**) With constant $\Delta_\nu T_s = 1 \times 10^{-4}$ and varying OSNR. (**b**) With constant OSNR = 10 dB and varying $\Delta_\nu T_s$. We assume a polarization-multiplexed QPSK signal with symbol rate $R_s = 28$ GBd and symbol energy $E_s = 1$. The lower the OSNR, the flatter the filter, as more symbols are required to filter out the additive noise contribution. On the other hand, the higher the phase noise, the more abrupt the filter edges, as the phase noise samples are less correlated

The rotated symbols $z_b[k]$ are subsequently applied to a minimum distance operator, and the quadratic distance $|d_b[k]|^2$ between the symbols before and after the decision operation is calculated

$$|d_b[k]|^2 = |z_b[k] - \lfloor z_b[k] \rfloor_D|^2, \tag{6.22}$$

where $\lfloor \cdot \rfloor_D$ denotes the minimum distance decision operation.

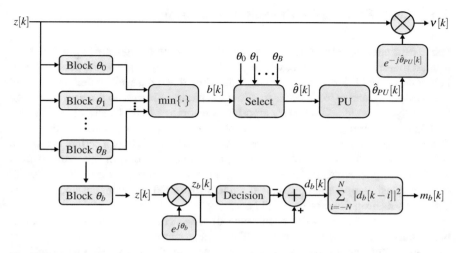

Fig. 6.10 BPS algorithm. Input signal $z[k]$ is rotated by B test phases θ_b. The rotated symbols $z_b[k]$ are subsequently applied to a minimum distance operator, and the quadratic distance $|d_b[k]|^2$ between the symbols before and after the decision operation is calculated. In order to mitigate the additive noise influence, the quadratic distances of $2N+1$ consecutive symbols rotated by the same test phase are added, producing signal $m_b[k]$. Estimated phase $\hat{\theta}[k]$ is selected as the $\theta_b[k]$ that minimizes $m_b[k]$. A PU is required to avoid discontinuities

In order to mitigate the additive noise influence on the phase recovery performance, the quadratic distances of $2N+1$ consecutive symbols rotated by the same test phase are added

$$m_b[k] = \sum_{i=-N}^{N} |d_b[k-i]|^2. \tag{6.23}$$

The choice of N is made based on the balance of phase noise power and OSNR. Finally, estimated phase $\hat{\theta}[k]$ is selected as the $\theta_b[k]$ that minimizes $m_b[k]$

$$b[k] = \min_b \{m_b[k]\}; \tag{6.24}$$

$$\hat{\theta}[k] = \theta_{b[k]}. \tag{6.25}$$

Estimated phases $\hat{\theta}[k]$ are limited to a span of p rad. For test phases calculated according to (6.20), the estimated phase $\hat{\theta}[k]$ has an excursion between $-p/2$ e $p/2$ rad. Therefore, a PU is required to avoid phase discontinuities.

6.3   Cycle Slips and Differential Encoding

Phase estimates generated in feedforward phase recovery algorithms are generally limited to $-\pi/P$ and π/P rad, where P is the ambiguity angle of the constellation. For square QAM formats, $P = 4$, indicating that phase estimates are limited between $-\pi/4$ and $\pi/4$. In order to allow the phase estimates to have an excursion from $-\infty$ to $+\infty$, phase recovery algorithms resort to a PU. The PU adds multiples of $\pi/2$ to the estimates generated by the phase recovery algorithm so that the phase difference between two consecutive estimates is always less than a certain threshold. Although necessary, the PU operation is also prone to errors. In scenarios of high additive and phase noise, the PU can add or subtract incorrect multiples of $\pi/2$, causing constellation rotations called cycle slips (CSs). Uncompensated CSs can generate catastrophic error bursts [17].

Figure 6.11 illustrates the occurrence of CSs. The blue curve is the original phase noise process to be estimated. The orange curve is the phase noise process estimated after feedforward phase recovery, e.g., by the Viterbi & Viterbi or BPS algorithms. As the original phase noise process is close to $\pi/4$, and the estimated phase noise process before the PU is limited between $-\pi/4$ and $\pi/4$, the estimated phase bounces between two values. The orange curve is much more noisy than the blue one because of AWGN. The green curve shows the estimated phases after the PU. At first, the PU succeeds in avoiding bouncing, by adding $\pi/2$ to the estimates with negative phases. However, because of the intense AWGN, it incorrectly adds $\pi/2$ to a sample with positive phase, causing a cycle slip. From this point on, the green curve still approaches the blue one, except for a shift of $\pi/2$.

There are several strategies to mitigate CSs. An interesting approach assigns the task of correcting CSs to the forward error correction (FEC) decoder [18]. Another approach is based on the periodic insertion of pilot symbols [19]. Pilots are very common in pragmatic DSP implementations, and the same pilots can be of help to several DSP stages. Error bursts caused by CSs are confined between pilots and can be eliminated by subsequent FEC schemes. The pilot rate must be tuned to avoid excessive bandwidth loss and still be able to compensate for cycle slips. However, the most deployed technique for avoiding error propagation caused by cycle slips is differential encoding (DE) and decoding.

The principle of DE is to encode two bits of information in the differential quadrant between consecutive symbols. Figure 6.12a illustrates the process of DE for the QPSK modulation format. If bits 00 are transmitted, two consecutive symbols have the same phase; if 01 are transmitted, there is a phase shift of $\pi/2$; if 11 are transmitted, there is a phase shift of π; if 10 are transmitted, there is a phase shift of $3\pi/2$. A cycle slip would cause a symbol error, but further error propagation is eliminated. Figure 6.12b illustrates DE for the 16-QAM format. In the first quadrant, bits are normally mapped using Gray mapping. Mapping in the other quadrants is a rotated version of that in the first quadrant. Just as in the case of QPSK, two bits are mapped to the differential quadrant between consecutive symbols.

Although simple and effective, DE has the drawback of increasing the bit error rate (BER) in the absence of CSs. Additive noise can cause an incorrect quadrant

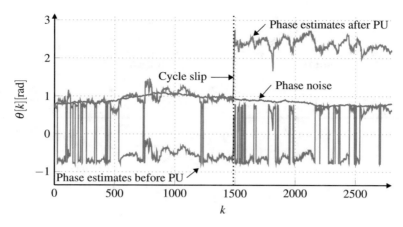

Fig. 6.11 Wiener phase noise (blue line) and its estimate before (orange line) and after (green line) the PU operation. Phase estimates before the PU are limited between $-\pi/4$ and $\pi/4$, supposing square QAM formats. The PU operation is necessary to allow phase estimates to have an excursion from $-\infty$ and ∞ and, thus, follow the trajectory of the original phase noise process. At first, the PU succeeds in avoiding phase bouncing, by adding $\pi/2$ to the estimates with negative phases. However, because of the intense AWGN, it incorrectly adds $\pi/2$ to a sample with positive phase, causing a cycle slip. From this point on, the green curve still approaches the green one, except for a shift of $\pi/2$

decision, generating two consecutive symbol errors after differential decoding. Figure 6.13 shows the BER *versus* SNR_b curves for the QPSK and 16-QAM formats, assuming an AWGN channel, for Gray mapping (solid lines) and for DE (dashed lines), obtained by Monte-Carlo simulation. The penalty for DE reaches approximately 0.6 dB for QPSK and 0.4 dB for 16-QAM at a BER of 10^{-3}.

6.4   Problems

1. Simulate the transmission of polarization-multiplexed signals. Assume Nyquist-shaped 16-QAM signals at 50 GBaud and an OSNR of 28 dB. For pulse shaping, use a root-raised cosine (RRC) shaping filter at 16 Sa/Symbol, with a roll-off factor $\beta^{RC} = 0.1$ and a span of 20 symbols.[1] Mix the signals in the two polarization orientations using a rotation angle of 10°. Introduce a frequency offset $\Delta f = 100$ MHz to both polarization orientations. At the receiver, after the optical front-end, apply a low-pass 10th order super-Gaussian filter with 25-GHz bandwidth to the signals, and sample them at 2 Sa/Symbol. Separate the polarization-multiplexed signals using the radius-

---

[1] In all simulation problems, transmit signals with at least 2^{16} symbols per polarization orientation. Consider optical modulation and coherent detection in the simulation models.

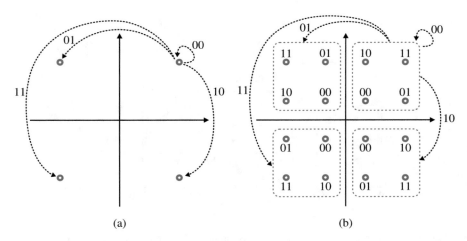

(a)   (b)

Fig. 6.12 DE of QPSK and 16-QAM formats. (**a**) For QPSK, the two bits define the quadrant transition between consecutive symbols. From an initial state, bits 00, 01, 10, and 11 indicate a phase transition of 0, $\pi/2$, $3\pi/2$, and π rad, respectively. (**b**) For 16-QAM, two out of the four bits define the transition between quadrants, while the other two bits are Gray-mapped within a quadrant. Note that the mapping within quadrants is rotated versions of the mapping in the first quadrant

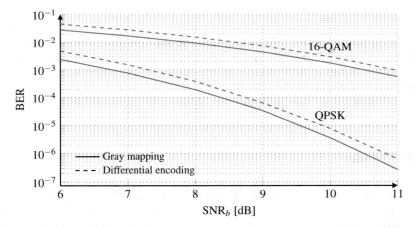

Fig. 6.13 BER *versus* SNR_b for the QPSK and 16-QAM formats, assuming an AWGN channel. Solid lines: Gray mapping. Dashed lines: DE. The curves were obtained by Monte-Carlo simulation

directed equalization (RDE) algorithm, with a pre-convergence stage using the constant modulus algorithm (CMA). Implement the equalizer with 11 taps. Discard the symbols generated during the convergence period of the adaptive equalizer. Estimate the frequency offset. Repeat the problem with $\Delta f = 250$

MHz, 500 MHz, 750 MHz, and 1 GHz. Compare the generated frequency offsets to their estimates.

2. Consider the scenario of Problem 1, with a frequency offset $\Delta f = 500$ MHz. Compensate for the frequency offset. Generate four scatter plots, one with the constellation of the last 1000 symbols of the signal before frequency recovery, and another with the constellation of the last 1000 symbols of the signal after frequency recovery, for both polarization orientations.

3. Plot a Wiener phase noise realization, assuming a sampling rate of 100 GSa/Symbol, and transmitter and LO lasers with 50 kHz each.

4. Simulate the transmission of a single-polarization signal. Assume Nyquist-shaped QPSK signals at 50 GBaud, and an OSNR of 28 dB. For pulse shaping, use the same parameters as in Problem 1. Consider at the transmitter a laser with 100-kHz linewidth. At the receiver, consider an ideal local oscillator with 0-Hz linewidth. Apply a low-pass 10th order super-Gaussian filter with 25-GHz bandwidth to the signals at the output of the optical front-end and sample the signals at 2 Sa/Symbol. After combining the in-phase and quadrature components, further downsample the signal to 1 Sa/Symbol. Estimate the phase noise process using the Viterbi & Viterbi algorithm with $N = 32$, the DD algorithm with $N = 40$, and the BPS algorithm with $N = 32$ and $B = 64$. Plot the downsampled version of the phase noise process produced by the transmitter laser, and their estimates produced by the Viterbi & Viterbi, decision-directed, and BPS algorithms. Repeat the problem with an OSNR of 10 dB. Discuss the results.

5. Simulate the transmission of polarization-multiplexed signals. Assume Nyquist-shaped QPSK signals at 50 GBaud and OSNR in the range of 12 dB to 22 dB. For pulse shaping, use the same parameters of Problem 1. Simulate lasers of 50-kHz linewidth, at both the transmit and the receive sides. Mix the signals of the two-polarization orientations using a rotation angle of 10°. Introduce a frequency offset of $\Delta f = 500$ MHz to both polarization orientations. Apply a low-pass 10th order super-Gaussian filter with 25 GHz bandwidth to the signals at the output of the optical front-end and sample the signals at 2 Sa/Symbol. Separate the polarization-multiplexed signals using the CMA, using a 11-tap equalizer. Then, perform frequency and phase recovery. For phase recovery, use the Viterbi & Viterbi algorithm with $N = 40$. Plot in the same figure the simulated and theoretical BER values. Synchronize the transmitted and received sequences before BER evaluation. Discuss the results.

6. Repeat Problem 5 using the DD algorithm with $N = 40$. Discard the first 3000 symbols generated after phase recovery.

7. Repeat Problem 5 using the BPS algorithm with $N = 40$ and $B = 32$.

8. Repeat Problem 5 considering Nyquist-shaped 16-QAM signals and the OSNR range of 16 dB to 26 dB. Separate the polarization-multiplexed signals using the RDE algorithm with a pre-convergence stage performed by the CMA. Use a 11-tap equalizer. For phase recovery, use the DD algorithm with $N = 60$. Discard the first 3000 symbols generated after phase recovery.

9. Repeat Problem 8 using the BPS algorithm with $N = 64$ and $B = 64$.

10. Simulate the transmission of polarization-multiplexed signals. Assume Nyquist-shaped QPSK signals at 50 GBaud and OSNR in the range of 5 dB to 15 dB. For pulse shaping, use the same parameters as in Problem 1. Consider initially ideal lasers of 0-Hz linewidth at both the transmit and receive sides. After the optical front-end, apply an RRC filter to the signals with the same parameters of the pulse shaping filter and sample the signals at 1 Sa/Symbol. Evaluate the BER for signals generated with Gray mapping and for signals generated with DE. Repeat the problem considering lasers of 50-kHz linewidth at both the transmit and receive sides, with Viterbi & Viterbi phase recovery using $N = 40$. Discuss the results.

11. Repeat Problem 10 considering 16-QAM signals and OSNRs in the interval of 8 dB to 18 dB. Use the BPS algorithm with $N = 64$ and $B = 64$ for phase recovery. Discuss the results.

6.5 Matlab/Octave Functions

6.5.1 Functions for Sect. 6.1

Matlab/Octave Code 6.1 Local oscillator carrier frequency shift

```
function [EOut] = LOFrequencyShift(EIn,Delta_f,SpS,Rs)
%%%%%%%%%%%%%%%%%%%%%%%%%%%%%%%%%%%%%%%%%%%%%%%%%%%%%%%%%%%%%%%%%%%%%%%%%%%%
% FOInsertion [EOut] = LOFrequencyShift(EIn,Delta_f,SpS,Rs)            %
%                                                                      %
%  This function shifts the carrier frequency of the local oscillator (LO)%
% laser by '-Delta_f'. To do so, a constant phase offset between       %
% neighboring samples calculated as                                    %
%                    DeltaTheta_f = -2*pi*Delta_f*T                     %
% where 'T' is the sampling period (in seconds) and 'Delta_f' is the   %
% carrier frequency shift (in Hz), is applied to the signal generated by%
% the LO laser, 'EIn'. This function can be used to insert a carrier   %
% frequency offset in the coherent detected signals. Assuming that the %
% carrier frequency of the transmitter and LO lasers are 'X Hz', shifting%
% the carrier frequency of the LO laser by '-Delta_f' produces a carrier%
% frequency offset of 'Delta_f' after coherent detection.              %
%                                                                      %
% Input:                                                               %
%   EIn     = Signal produced by the LO laser (e.g. produced by  the   %
%             function 'Laser'). For transmission in single pol.       %
%             orientation, 'EIn' must be a column vector. For transmission%
%             with pol. multiplexing, 'EIn' must be a matrix with two  %
%             column-oriented vectors, where each column vector represents%
%             the signal of a pol. orientation;                        %
%   Delta_f = Carrier frequency shift (in Hz) to be applied in the signal%
%             'EIn';                                                    %
%   SpS     = Number of samples per symbol in the oversampled signal   %
%             (transmitted signal / received signal before sampling);  %
%   Rs      = Symbol rate in symbols/second;                           %
%                                                                      %
% Output:                                                              %
%   EOut = Signal generated after applying the frequency shift in 'Ein';%
%                                                                      %
% This function is part of the book Digital Coherent Optical Systems;  %
% Darli A. A. Mello and Fabio A. Barbosa;                              %
```

```
%%%%%%%%%%%%%%%%%%%%%%%%%%%%%%%%%%%%%%%%%%%%%%%%%%%%%%%%%%%%%%%%%%%%%%%%%%%%%

    % Period between samples at the (oversampled) transmitted signal:
    T = 1/(SpS*Rs);

    % Phase offset between consecutive samples due to 'Delta_f':
    DeltaTheta_f = -2*pi*Delta_f*T;

    % Inserting the effects of frequency offset into signal 'EIn':
    k    = repmat((0:size(EIn,1)-1).',1,size(EIn,2));
    EOut = EIn.*exp(1i*k*DeltaTheta_f);
end
```

Matlab/Octave Code 6.2   Frequency recovery algorithm

```
function [z,varargout] = FRecovery(y,Rs,FEstimate)
%%%%%%%%%%%%%%%%%%%%%%%%%%%%%%%%%%%%%%%%%%%%%%%%%%%%%%%%%%%%%%%%%%%%%%%%%%%%
% FOCOMPENSATION [z] = FRecovery(y,Rs,FEstimate)                         %
%                                                                        %
%   This function performs frequency recovery in the signal 'y'. Using the %
% spectal analysis of the signal 'y' raised to 4-th power an estimate    %
% 'Delta_f' of the carrier frequency offset is generated. Then, the      %
% frequency recovery is performed as                                     %
%                     z = y.*exp(-1i*k*2*pi*Delta_f*Ts)                   %
% where 'Ts' is the symbol period and 'k' is the temporal index of the   %
% symbols;                                                               %
%                                                                        %
% Input:                                                                  %
%   y         = Signal in which the frequency recovery will be performed. %
%               Signal 'y' must be obtained at 1 sample per symbol. For   %
%               transmission in single pol. orientation, 'y' must be a    %
%               column vector. For transmission with pol. multiplexing,   %
%               'y' must be a matrix with two column-oriented vectors,    %
%               where each column vector corresponds to the signal of one %
%               pol. orientation;                                        %
%   Rs        = Symbol rate in symbols/second;                           %
%   FEstimate = Flag to enable ('true') or disable ('false') the estimated%
%               carrier frequency offset value as an output of the function;%
%                                                                        %
% Output:                                                                 %
%   z         = Signal produced after compensating for the frequency      %
%               offset present on 'y';                                    %
%   varargout = Estimated carrier frequency offsert in Hz;               %
%                                                                        %
% This function is part of the book Digital Coherent Optical Systems;    %
% Darli A. A. Mello and Fabio A. Barbosa;                                %
%%%%%%%%%%%%%%%%%%%%%%%%%%%%%%%%%%%%%%%%%%%%%%%%%%%%%%%%%%%%%%%%%%%%%%%%%%%%

    % Frequency vector considering 1 sample per symbol and symbol period:
    f = (-1/2+1/length(y):1/length(y):1/2)*Rs ; Ts = 1/Rs;

    % Obtaining the absolute value of the spectrum of the signal^4:
    SignalSpectrum = fftshift(abs(fft(y(:,1).^4)));

    % Obtaining the frequency offset:
    Delta_f = (1/4)*f(SignalSpectrum == max(SignalSpectrum));

    % Compensating for the frequency offset:
    k = repmat((0:length(y)-1).',1,size(y,2));
    z = y.*exp(-1i*2*pi*Delta_f*Ts*k);

    % Estimated carrier frequency offset as an output of the function:
    if FEstimate
        varargout{1} = Delta_f;
    end
end
```

6.5.2   Functions for Sect. 6.2

Matlab/Octave Code 6.3 ML filter for Viterbi & Viterbi algorithm

```
function [wML] = MLFilterViterbi(M,Delta_nu,Rs,OSNRdB,Es,NPol,N)
%%%%%%%%%%%%%%%%%%%%%%%%%%%%%%%%%%%%%%%%%%%%%%%%%%%%%%%%%%%%%%%%%%%%%%%%%%
% MLFILTER [wML] = MLFilterViterbi(M,Delta_nu,Rs,OSNRdB,Es,NPol,N)      %
%                                                                       %
%  This function calculates the maximum likelihood (ML) filter for the  %
% Viterbi & Viterbi algorithm, which is calculated as                   %
%                         wML = (1^T C^-1)^T                             %
% where (.)^T indicates transpose and C is a covariance matrix, which   %
% depends on the signal-to-noise ratio and on the phase noise magnitude.%
%                                                                       %
% Input:                                                                %
%   M        = Modulation order of the M-PSK modulation format;         %
%   Delta_nu = Sum of transm. and local oscillator laser linewidths in Hz;%
%   Rs       = Symbol rate in symbols/second;                           %
%   OSNRdB   = Channel OSNR in dB;                                      %
%   Es       = Symbol energy (per pol. orientation) in W;               %
%   NPol     = Number of pol. orientations used;                        %
%   N        = Number of past and future symbols used in the Viterbi &  %
%              Viterbi algorithm for phase noise estimates. The block   %
%              length is then L = 2*N+1;                                 %
%                                                                       %
% Output:                                                               %
%   wML = Maximum likelihood filter to be used in the Viterbi & Viterbi %
%         algorithm for phase noise estimation. wML is a column vector; %
%                                                                       %
% This function is part of the book Digital Coherent Optical Systems;   %
% Darli A. A. Mello and Fabio A. Barbosa;                               %
%%%%%%%%%%%%%%%%%%%%%%%%%%%%%%%%%%%%%%%%%%%%%%%%%%%%%%%%%%%%%%%%%%%%%%%%%%

    % Block length for the Viterbi & Viterbi algorithm, symbol period, and
    % phase noise variance:
    L = 2*N+1 ; Ts = 1/Rs ; Sigma_DeltaTheta2 = 2*pi*Delta_nu*Ts;

    % Additive noise variance:
    SNRLin=10^(OSNRdB/10)*(2*12.5e9)/(NPol*Rs) ; Sigma_eta2= Es/(2*SNRLin);

    % K matrix:
    KAux = zeros(N) ; K    = zeros(L);
    for i = 0:N
        for ii = 0:N
            KAux(i+1,ii+1) = min(i,ii);
        end
    end
    K(1:N+1,1:N+1) = rot90(KAux(1:N+1,1:N+1),2);
    K(N+1:L,N+1:L) = KAux(1:N+1,1:N+1);

    % Identity matrix:
    I = eye(L);

    % Obtaining the covariance matrix:
    C = Es^M*M^2*Sigma_DeltaTheta2*K + Es^(M-1)*M^2*Sigma_eta2*I;

    % Filter coefficients:
    wML = (ones(L,1)'/(C)).' ; wML = wML/max(wML);
end
```

Matlab/Octave Code 6.4 Viterbi & Viterbi algorithm

```
function [v,varargout] = ViterbiCPR(z,Delta_nu,Rs,OSNRdB,Es,NPol,M,...
    ParamViterbi)
%%%%%%%%%%%%%%%%%%%%%%%%%%%%%%%%%%%%%%%%%%%%%%%%%%%%%%%%%%%%%%%%%%%%%%%%%%%%%%
% VITERBICPR [v,varargout] = ViterbiCPR(z,Delta_nu,Rs,OSNRdB,Es,NPol,M,...%
%                           ParamViterbi)                                 %
%                                                                         %
%  This function performs phase recovery of signals with M-PSK modulation %
% formats using the Viterbi & Viterbi algorithm. The signal 'z' is first  %
% raised to M-th power and then ML filtering is performed before phase    %
% estimate. The phase estimates are applied to a phase unwrapper and,     %
% finally, the phase noise is compensated.                                %
%                                                                         %
% Input:                                                                  %
%   z           = Signal in which phase recovery will be performed. For   %
%                 transmission in single pol. orientation, 'z' must be a  %
%                 column vector. For transmittion with pol. multiplexing, %
%                 'z' must be a matrix with two column-oriented vectors,  %
%                 where each column vector corresponds to the signal of   %
%                 one pol. orientation. Signal 'z' must be obtained at 1  %
%                 sample per symbol. Signal 'z' must also be normalized   %
%                 to unitary power;                                       %
%   Delta_nu    = Sum of the transmitter and local oscillator laser       %
%                 linewidths in Hz;                                       %
%   Rs          = Symbol rate in symbols/second;                          %
%   OSNRdB      = Channel OSNR in dB;                                     %
%   Es          = Symbol energy (per pol. orientation) in W;              %
%   NPol        = Number of pol. orientations used;                       %
%   M           = Modulation order of the M-PSK modulation format;        %
%   ParamViterbi = Struct that specifies parameters of the Viterbi        %
%                 & Viterbi algorithm:                                    %
%       -ParamViterbi.N: Number of past and future symbols used in the    %
%        Viterbi & Viterbi algorithm for phase noise estimate. The block  %
%        length is then 'L = 2*N+1';                                      %
%       -ParamViterbi.PEstimate: Flag to enable ('true') or disable       %
%        ('false') the estimated phase noise as an output of the function;%
%                                                                         %
% Output:                                                                 %
%   v = Signal produced after compensating for the phase noise present on %
%       'z';                                                              %
%   varargout: When the flag 'ParamViterbi.PEstimate' is 'true', the      %
%              estimated phase noise is also an output of the function;    %
%                                                                         %
% This function is part of the book Digital Coherent Optical Systems;     %
% Darli A. A. Mello and Fabio A. Barbosa;                                 %
%%%%%%%%%%%%%%%%%%%%%%%%%%%%%%%%%%%%%%%%%%%%%%%%%%%%%%%%%%%%%%%%%%%%%%%%%%%%%%

    % Block length for the phase estimation:
    N = ParamViterbi.N ; L = 2*N+1;

    % ML filter:
    wML = MLFilterViterbi(M,Delta_nu,Rs,OSNRdB,Es,NPol,N);

    % Initializing the vector of phase estimates:
    ThetaML = zeros(size(z,1),NPol);

    % Phase noise estimation:
    for Pol = 1:NPol
        % Input blocks:
        zBlocks = [zeros(floor(L/2),1) ; z(:,Pol) ; zeros(floor(L/2),1)];
        zBlocks=convmtx(zBlocks.',L); zBlocks=flipud(zBlocks(:,L:end-L+1));

        % Note that each column of zBlocks is used independently to
        % generate phase estimates:
```

```
            ThetaML(:,Pol) = (1/M)*angle(wML.'*(zBlocks.^M))-pi/M;
    end
    clearvars zBlocks;

    % Vector of phase estimates after phase unwrapping:
    ThetaPU = zeros(size(ThetaML,1),NPol);

    % Initial 'previous phase' for unwrapping operation:
    ThetaPrev = zeros(1,NPol);

    % Phase unwrapping:
    for i = 1:size(ThetaML,1)
        % Phase unwrapper:
        n           = floor(1/2 + (ThetaPrev - ThetaML(i,:))/(2*pi/M));
        ThetaPU(i,:) = ThetaML(i,:) + n*(2*pi/M); ThetaPrev = ThetaPU(i,:);
    end

    % Phase noise compensation:
    v = z.*exp(-1i*ThetaPU);

    % Estimated phase noise as an output of the function:
    if ParamViterbi.PEstimate
        varargout{1} = ThetaPU;
    end
end
```

Matlab/Octave Code 6.5 ML filter for DD phase recovery algorithm

```
function [wML] = MLFilterDD(Delta_nu,Rs,OSNRdB,Es,NPol,N)
%%%%%%%%%%%%%%%%%%%%%%%%%%%%%%%%%%%%%%%%%%%%%%%%%%%%%%%%%%%%%%%%%%%%%%%%%%%
% MLFilterDD [wML] = MLFilterDD(Delta_nu,Rs,OSNRdB,Es,NPol,N)          %
%                                                                      %
%  This function calculates the maximum likelihood (ML) filter for the  %
% Decision-Directed algorithm, which is calculated as                  %
%                     wML = (1^T C^-1)^T                                %
% where (.)^T indicates transpose and C is a covariance matrix, which   %
% depends on the signal-to-noise ratio and on the phase noise magnitude. %
%                                                                      %
% Input:                                                               %
%   Delta_nu = Sum of the transmitter and local oscillator laser       %
%              linewidths in Hz;                                       %
%   Rs       = Symbol rate in Symbols/second;                          %
%   OSNRdB   = Channel OSNR in dB;                                     %
%   Es       = Symbol energy (per po. orientation) in W;               %
%   NPol     = Number of pol. orientations used;                       %
%   N        = Number of symbols used in the Decision-Directed algorithm %
%              for phase noise estimation;                             %
%                                                                      %
% Output:                                                              %
%   wML = Maximum likelihood filter to be used in the Decision-Directed %
%         algorithm for phase noise estimation. wML is a column vector; %
%                                                                      %
% This function is part of the book Digital Coherent Optical Systems;   %
% Darli A. A. Mello and Fabio A. Barbosa;                              %
%%%%%%%%%%%%%%%%%%%%%%%%%%%%%%%%%%%%%%%%%%%%%%%%%%%%%%%%%%%%%%%%%%%%%%%%%%%

    % Symbol period and phase noise variance:
    Ts = 1/Rs ; Sigma_DeltaTheta = 2*pi*Delta_nu*Ts;

    % Additive noise variance:
    SNRLin=10^(OSNRdB/10)*(2*12.5e9)/(NPol*Rs) ; Sigma_eta = Es/(2*SNRLin);

    % K matrix:
    K = zeros(N);
    for i = 0:N-1
        for j = 0:N-1
```

```
             K(i+1,j+1) = min(i,j);
         end
    end

    % Identity matrix:
    I = eye(N);

    % Obtaining the covariance matrix:
    C = Es^2*Sigma_DeltaTheta*K + Es*Sigma_eta*I;

    % Filter coefficients:
    wML = (ones(N,1)'/(C)).' ; wML = wML/max(wML);
end
```

Matlab/Octave Code 6.6 DD phase recovery algorithm

```
function [s_Decided,varargout] = DDCPR(z,ModFormat,Delta_nu,Rs,OSNRdB,Es...
    ,NPol,ParamDD)
%%%%%%%%%%%%%%%%%%%%%%%%%%%%%%%%%%%%%%%%%%%%%%%%%%%%%%%%%%%%%%%%%%%%%%%%%%%%%
% DDCPR [s_Decided,varargout] = DDCPR(z,ModFormat,Delta_nu,Rs,OSNRdB,Es...%
%                              ,NPol,ParamDD)                            %
%                                                                        %
%  This function performs phase recovery in signal 'z' with modulation   %
% format defined by 'ModFormat' using the Decision-Directed algorithm. The%
% algorithm uses 'ParamDD.N' previous symbol decisions to estimate the   %
% phase noise. In addition, a maximum likelihood (ML) filter is used prior%
% to generate phase estimates.                                           %
%                                                                        %
% Input:                                                                 %
%   z          = Signal in which phase recovery will be performed. For   %
%                transmission in single pol. orientation, 'z' must be a  %
%                column vector. For transmission with pol. multiplexing,'z'%
%                must be a matrix with two column vectors, where each     %
%                column vector corresponds to the signal of one pol.      %
%                orientation. Signal 'z' must be obtained at 1 sample per %
%                symbol. Signal 'z' must also be normalized to unitary power;%
%   ModFormat  = Modulation format of the signal 'z';                    %
%   Delta_nu   = Sum of the transmitter and local oscillator laser       %
%                linewidths in Hz;                                       %
%   Rs         = Symbol rate in symbols/second;                          %
%   OSNRdB     = Channel OSNR in dB;                                     %
%   Es         = Symbol energy (per pol. orientation) in W;              %
%   NPol       = Number of pol. orientations used;                       %
%   ParamDD    = Struct that specifies parameters of the decision-directed %
%                phase recovery algorithm:                               %
%      -ParamDD.N: Number of past symbols used in the algorithm for phase%
%       noise estimation;                                                %
%      -ParamDD.OutSymb: Flag to enable ('true') or disable ('false') the%
%       symbols prior to decision as output;                             %
%      -ParamDD.PEstimate: Flag to enable ('true') or disable ('false')  %
%       the estimated phase noise as an output of the function;          %
%                                                                        %
% Output:                                                                %
%   s_Decided = Sequence of (decided) symbols produced after compensating %
%               for the phase noise present on 'z';                      %
%   varargout: (1) When the flag 'ParamDD.OutSymb' is 'true', the symbols %
%         obtained prior to decision, 'v', are an output of the function; %
%              (2) When the flag 'ParamDD.PEstimate' is 'true', the      %
%         estimated phase noise is also an output of the function;       %
% Note:-Both the flags 'ParamDD.OutSymb' and 'ParamDD.PEstimate' can be  %
%    set 'true' at the same execution. In this case the function has 3    %
%    outputs:                                                            %
%                    [s_Decided,v,PEstimate] = DDCPR(....)               %
%    - If only the flag 'ParamDD.OutSymb' is 'true',the outputs are:     %
%                    [s_Decided,v] = DDCPR(....)                         %
%    - If only the flag 'PEstimate' is 'true', the outputs are:          %
```

```
%                      [s_Decided,PEstimate] = DDCPR(....)                    %
%                                                                             %
% This function is part of the book Digital Coherent Optical Systems;         %
% Darli A. A. Mello and Fabio A. Barbosa;                                     %
%%%%%%%%%%%%%%%%%%%%%%%%%%%%%%%%%%%%%%%%%%%%%%%%%%%%%%%%%%%%%%%%%%%%%%%%%%%%%%%%

    % Vectors of phase estimates and output symbols:
    Theta = zeros(size(z,1),NPol) ; s_Decided = zeros(size(z,1),NPol);

    % Initializing the buffer of symbols used for phase recovery:
    N = ParamDD.N ; b = ones(N,NPol);

    % ML filter:
    wML = MLFilterDD(Delta_nu,Rs,OSNRdB,Es,NPol,N);

    % Phase noise estimates:
    for i = 1:length(z)
        % Obtaining the phase of the previous received symbol:
        Theta(i,:) = angle(wML'*b);

        % Compensating for the phase of the received symbols:
        v = z(i,:).*exp(-1i*Theta(i,:));

        % Deciding the symbols after phase recovery:
        s_Decided(i,:) = Decision(v,ModFormat,false);

        % Updating the buffer b = z*s_Decided:
        b = circshift(b,1,1);
        b(1,:) = z(i,:).*conj(s_Decided(i,:))...
            ./abs(z(i,:).*conj(s_Decided(i,:)));
    end

    % Symbols prior to decision:
    AuxOut = 1;
    if ParamDD.OutSymb
        varargout{AuxOut} = z.*exp(-1i*Theta) ; AuxOut = AuxOut + 1;
    end

    % Estimated phase noise as an output of the function:
    if ParamDD.PEstimate
        varargout{AuxOut} = unwrap(Theta);
    end
end
```

Matlab/Octave Code 6.7 BPS algorithm

```
function [v,varargout] = BPS(z,ModFormat,NPol,ParamBPS)
%%%%%%%%%%%%%%%%%%%%%%%%%%%%%%%%%%%%%%%%%%%%%%%%%%%%%%%%%%%%%%%%%%%%%%%%%%%%%%%%
% BPS [v,varargout] = BPS(z,ModFormat,NPol,ParamBPS)                          %
%                                                                             %
%  This function performs phase recovery in signal 'z' with modulation        %
% format defined by 'ModFormat' using the blind phase search (BPS)            %
% algorithm. The algorithm uses 'L = 2*ParamBPS.N+1' symbols and             %
% 'ParamBPS.B' test rotations to generate phase noise estimates.             %
%                                                                             %
% Input:                                                                      %
%   z        = Signal in which phase recovery will be performed. For          %
%                transmission in single pol. orientation, 'z' must be a       %
%                column vector. For transmission with pol. multiplexing,      %
%                'z' must be a matrix with two column vectors, where each     %
%                column vector corresponds to the signal of one pol.          %
%                orientation. Signal 'z' must be obtained at 1 sample per     %
%                symbol. Signal 'z' must also be normalized to unitary power; %
%   ModFormat = Modulation format of the signal 'z';                         %
%   NPol      = Number of pol. orientations used;                            %
%   ParamBPS  = Struct that specifies parameters of the BPS algorithm:        %
```

```
%          -ParamBPS.N: Number of 'past' and 'future' symbols used in the    %
%           BPS algorithm for phase noise estimation. The total number of    %
%           symbols is then L = 2*N+1;                                        %
%          -ParamBPS.B: Number of test rotations;                            %
%          -ParamBPS.PEstimate: Flag to enable ('true') or disable ('false')%
%           the estimated phase noise as an output of the function;          %
%                                                                            %
% Output:                                                                    %
%   v       = Signal produced after compensating for the phase noise present%
%             on 'z';                                                        %
%   varargout: When the flag 'ParamBPS.PEstimate' is 'true', the estimated%
%              phase noise is also an output of the function;               %
%                                                                            %
% This function is part of the book Digital Coherent Optical Systems;        %
% Darli A. A. Mello and Fabio A. Barbosa;                                    %
%%%%%%%%%%%%%%%%%%%%%%%%%%%%%%%%%%%%%%%%%%%%%%%%%%%%%%%%%%%%%%%%%%%%%%%%%%%%%%%%

    % Number of test rotations and ambiguity angle; Total BPS block length:
    B = ParamBPS.B ; p = pi/2 ; N = ParamBPS.N ; L = 2*N+1;

    % Creating a vector of test carrier phase angles:
    b = -B/2:1:B/2-1 ; ThetaTest = p*b/B;

    % Creating a matrix of test carrier phase angles:
    ThetaTestMatrix = repmat(exp(-1j*ThetaTest),L,1);
    if NPol == 2
        ThetaTestMatrix = cat(3,ThetaTestMatrix,ThetaTestMatrix);
    end

    % Input blocks:
    % V-pol. orientation:
    zB_V = [zeros(floor(L/2),1); z(:,1) ; zeros(floor(L/2),1)];
    zB_V = convmtx(zB_V.',L)    ; zB_V = flipud(zB_V(:,L:end-L+1));
    if NPol == 2
        % H-pol. orientation:
        zB_H = [zeros(floor(L/2),1)  ; z(:,2) ; zeros(floor(L/2),1)];
        zB_H = convmtx(zB_H.',L)     ; zB_H = flipud(zB_H(:,L:end-L+1));
        zBlocks   = cat(3,zB_V,zB_H) ; clearvars zBlocks_V zBlocks_H;
    else
        zBlocks = zB_V; clearvars zBlock_V;
    end

    % Vector of phase estimates and initial phase for the PU:
    ThetaPU = zeros(size(zBlocks,2),NPol) ; ThetaPrev = zeros(1,NPol);

    % Phase noise estimates:
    for i = 1:size(zBlocks,2)
        % Applying the test phase angles to the symbols:
        zRot = repmat(zBlocks(:,i,:),1,B,1).*ThetaTestMatrix;

        % Decision of the rotated symbols:
        zRot_Decided = Decision(zRot,ModFormat,false);

        % Intermidiate signal to be minimized:
        m = sum(abs(zRot-zRot_Decided).^2,1);

        % Estimating the phase noise as the angle that minimizes 'm':
        [~,im] = min(m,[],2) ; Theta  = reshape(ThetaTest(im),1,NPol);

        % Applying the phase unwrapper to the estimated phase angles:
        ThetaPU(i,:) = Theta + floor(0.5 - (Theta-ThetaPrev)./(p)).*(p);

        % Updating the previous phase variable:
        ThetaPrev = ThetaPU(i,:);
    end

    % Compensating for the phase noise:
```

```
    v = z.*exp(-1i*ThetaPU);

    % Estimated phase noise as an output of the function:
    if ParamBPS.PEstimate
        varargout{1} = ThetaPU;
    end
end
```

6.5.3   Functions for Sect. 6.3

Matlab/Octave Code 6.8 QPSK and 16-QAM symbol generation considering differential encoding

```
function [Bits,x] = SymbolGeneration_DiffEnc(ModFormat,NSymb)
%%%%%%%%%%%%%%%%%%%%%%%%%%%%%%%%%%%%%%%%%%%%%%%%%%%%%%%%%%%%%%%%%%%%%%%%%
% SymbolGeneration_DiffEnc [Txbits,x] = SymbolGeneration_DiffEnc(...   %
%                              ModFormat,NSymb)                        %
%                                                                     %
%  This function generates a sequence of bits 'Bits' with uniform     %
% distribution and, considering differential encoding, generates a    %
% sequence of symbols 'x' according to the modulation format 'ModFormat'. %
% The length of 'x' is given by 'NSymb'. The symbols of 'x' follow the %
% unitary power constellation associated with 'ModFormat'. For         %
% transmission with pol. multiplexing, this function must be called twice.%
%                                                                     %
% Input:                                                              %
%   ModFormat  = Modulation format: 'QPSK' or '16QAM';                %
%   NSymb      = Number of symbols to be transmitted;                 %
%                                                                     %
% Output:                                                             %
%   Bits     = Bits to be transmitted (column vector);               %
%   x        = Sequence of symbols (column vector);                  %
%                                                                     %
% This function is part of the book Digital Coherent Optical Systems;  %
% Darli A. A. Mello and Fabio A. Barbosa;                             %
%%%%%%%%%%%%%%%%%%%%%%%%%%%%%%%%%%%%%%%%%%%%%%%%%%%%%%%%%%%%%%%%%%%%%%%%%

    switch ModFormat
        case {'QPSK'}
            ModBits = 2;
        case '16QAM'
            ModBits = 4;
        otherwise
            error('The supported modulation formats are QPSK and 16-QAM;');
    end

    % Generating bits to be transmitted:
    NBits = NSymb*ModBits; Bits = randi([0,1],NBits,1);

    switch ModFormat
        case 'QPSK'
            % Input bits:
            QBits1 = Bits(1:ModBits:NBits); QBits2 = Bits(2:ModBits:NBits);

            % Defining the quadrant:
            Quad = ((~QBits1&~QBits2)*1            + ...
                    (~QBits1& QBits2)*exp(1i*pi/2) + ...
                    ( QBits1&~QBits2)*exp(3i*pi/2) + ...
                    ( QBits1& QBits2)*exp(1i*pi));
```

```
                % Initial quadrant:
                QuadPrev = 1;

                % Modulation:
                x = NaN(1,length(Quad));
                for i = 1:length(x)
                    x(i)     = QuadPrev*Quad(i)*(1+1i);
                    QuadPrev = QuadPrev*Quad(i);
                end
                x = (1/sqrt(2))*x.';

        case '16QAM'
                % Bits that define the quadrant:
                QBits1 = Bits(1:ModBits:NBits); QBits2 = Bits(2:ModBits:NBits);

                % Bits that define the symbols inside the quadrants:
                InQuadBits1 = Bits(3:ModBits:NBits);
                InQuadBits2 = Bits(4:ModBits:NBits);

                % Defining the quadrant:
                Quad = ((~QBits1&~QBits2)*1          + ...
                        (~QBits1& QBits2)*exp(1i*pi/2) + ...
                        ( QBits1&~QBits2)*exp(3i*pi/2) + ...
                        ( QBits1& QBits2)*exp(1i*pi));

                % Defining the symbol inside the quadrants:
                InQuadI     = 2*InQuadBits1 + 1; InQuadQ = 2*InQuadBits2 + 1;
                InQuadSymbol = InQuadI + 1i*InQuadQ;

                % Initial quadrant:
                QuadPrev = 1;

                % Modulation:
                x = NaN(1,length(InQuadBits2));
                for i = 1:length(x)
                   x(i)     = QuadPrev*Quad(i)*InQuadSymbol(i);
                   QuadPrev = QuadPrev*Quad(i);
                end
                x = ((1/sqrt(10))*x).';
    end
end
```

Matlab/Octave Code 6.9 Decision of QPSK and 16-QAM symbols considering differential decoding

```
function [Decided] = Decision_DiffEnc(r,ModFormat)
%%%%%%%%%%%%%%%%%%%%%%%%%%%%%%%%%%%%%%%%%%%%%%%%%%%%%%%%%%%%%%%%%%%%%%%%%%%%
% Decision_DiffEnc [Decided] = Decision_DiffEnc(r,ModFormat)             %
%                                                                        %
%   This function performs differential decoding on the received sequence %
% 'r' according to the modulation format 'ModFormat'. As output, the     %
% function generates the binary sequence associated with the sequence 'r'.%
%                                                                        %
% Input:                                                                 %
%   r        = Received signal (one pol. orientation - column vector)    %
%              normalized to unitary power (column vector);              %
%   ModFormat = Modulation format: 'QPSK' or '16-QAM';                   %
%                                                                        %
% Output:                                                                %
%   Decided = Binary sequence associated with sequence 'r' (column vector)%
%                                                                        %
% This function is part of the book Digital Coherent Optical Systems;    %
% Darli A. A. Mello and Fabio A. Barbosa;                                %
%%%%%%%%%%%%%%%%%%%%%%%%%%%%%%%%%%%%%%%%%%%%%%%%%%%%%%%%%%%%%%%%%%%%%%%%%%%%

    switch ModFormat
```

```
    case 'QPSK'
        ModBits = 2;
    case '16QAM'
        ModBits = 4;
    otherwise
        error('The supported modulation formats are QPSK and 16-QAM;');
end

% Decision:
QPrev = 0;
Decided  = NaN(length(r),ModBits);
switch ModFormat
    case 'QPSK'
        % Decision regions for the in-phase component:
        R1 = real(r) >= 0 ; R2 = real(r) < 0;
        % Decision regions for the quadrature component:
        R3 = imag(r) >= 0 ; R4 = imag(r) < 0;

        % Defining the quadrant:
        Q(R1&R3) = 0 ; Q(R2&R3) = 1 ; Q(R2&R4) = 2 ; Q(R1&R4) = 3;

        % Binary sequence:
        for i = 1:length(r)
            QRx = (Q(i) - QPrev);
            % Bits defining the quadrant (and consequently the symbol):
            switch QRx
                case 0
                    Decided(i,1) = 0; Decided(i,2) = 0;
                case {1,-3}
                    Decided(i,1) = 0; Decided(i,2) = 1;
                case {3,-1}
                    Decided(i,1) = 1; Decided(i,2) = 0;
                case {2,-2}
                    Decided(i,1) = 1; Decided(i,2) = 1;
            end
            QPrev = Q(i);
        end
    case '16QAM'
        % Applying the decision regions to the real axis:
        R1 = real(r) >= 2/sqrt(10); R2 = real(r) >= 0;
        R3 = real(r) < 0          ; R4 = real(r) <= -2/sqrt(10);
        % Applying the decision regions to the imaginary axis:
        R5 = imag(r) >= 2/sqrt(10); R6 = imag(r) >= 0;
        R7 = imag(r) < 0          ; R8 = imag(r) <= -2/sqrt(10);

        % Defining the quadrant:
        Q(R1&R5 | R1&R6&~R5 | ~R1&R2&R5 | ~R1&R2&R6&~R5) = 0;
        Q(~R4&R3&R5 | ~R4&R3&R6&~R5 | R4&R5 | R4&R6&~R5) = 1;
        Q(R4&R7&~R8 | R4&R8 | ~R4&R3&R7&~R8 | ~R4&R3&R8) = 2;
        Q(~R1&R2&R7&~R8 | ~R1&R2&R8 | R1&R7&~R8 | R1&R8) = 3;

        % Defining the symbol inside the quadrants:
        S(~R1&R2&~R5&R6|~R1&R2&~R8&R7|~R4&R3&~R5&R6|~R4&R3&~R8&R7) = 0;
        S(~R1&R2&R5 | R4&R6&~R5 | ~R4&R3&R8 | R1&R7&~R8)          = 1;
        S(R1&R6&~R5 | R4&R7&~R8 | ~R4&R3&R5 | ~R1&R2&R8)          = 2;
        S(R1&R5     | R1&R8     | R4&R5     | R4&R8)              = 3;

        % Received binary sequence:
        for i = 1:length(r)
            % Bits defining the quadrant:
            QRx = (Q(i) - QPrev);
            switch QRx
                case 0
                    Decided(i,1) = 0; Decided(i,2) = 0;
                case {1,-3}
                    Decided(i,1) = 0; Decided(i,2) = 1;
                case {3,-1}
```

```
                                  Decided(i,1) = 1; Decided(i,2) = 0;
                        case {2,-2}
                                  Decided(i,1) = 1; Decided(i,2) = 1;
                    end
                    QPrev = Q(i);
                    % Bits defining the symbol inside the quadrant:
                    switch S(i)
                        case 0
                                  Decided(i,3) = 0; Decided(i,4) = 0;
                        case 1
                                  Decided(i,3) = 0; Decided(i,4) = 1;
                        case 2
                                  Decided(i,3) = 1; Decided(i,4) = 0;
                        case 3
                                  Decided(i,3) = 1; Decided(i,4) = 1;
                    end
                end
        end
    end
    % Obtaining the vector of bits:
    Decided = reshape(Decided',1,length(r)*ModBits)';
end
```

Matlab/Octave Code 6.10 BER evaluation for differential decoding

```
function [BER] = ErrorCounting_DiffEnc(Rx,TxBits,ModFormat,NPol)
%%%%%%%%%%%%%%%%%%%%%%%%%%%%%%%%%%%%%%%%%%%%%%%%%%%%%%%%%%%%%%%%%%%%%%%%%%
% ErrorCounting [BER] = ErrorCounting_DiffEnc(Rx,Tx,ModFormat,...       %
%                                   MapperMode,NPol)                    %
%                                                                       %
%    This function evaluates the BER using the transmitted bit sequence %
% 'TxBits' and the received symbol sequence 'Rx', considering differential%
% encoding. The received sequence of bits is generated after applying   %
% differential decoding on the sequence 'Rx'. The sequence 'Rx' and the %
% sequence of transmitted symbols associated with the bit sequence      %
% 'TxBits' (i.e. 'Tx') must be synchronized prior to error evaluation. The%
% sequence 'Rx' must be normalized to unitary power.                    %
%                                                                       %
% Input:                                                                %
%    Rx       = Sequence of received symbols normalized to unitary power*.%
%    TxBits   = Sequence of transmitted bits associated with sequence 'Tx'*;%
%    ModFormat = Modulation format: 'QPSK' or '16-QAM';                 %
%    NPol     = Number of pol. orientations used;                       %
%   *Note: For transmission in single pol. orientation, 'TxBits' and 'Rx' %
%          must be column vectors. For transmission with pol. multiplexing,%
%          'TxBits' and 'Rx' must be matrices with two column vectors,  %
%          where each column vector corresponds to the signal of one pol. %
%          orientation.                                                 %
%                                                                       %
% Output:                                                               %
%    BER = Struct with the evaluated BER per pol. orientation. If NPol == 1%
%          the struct BER has only the field BER.V. If NPol == 2, the   %
%          struct BER has the filed BER.V and BER.H;                    %
%                                                                       %
% This function is part of the book Digital Coherent Optical Systems;   %
% Darli A. A. Mello and Fabio A. Barbosa;                               %
%%%%%%%%%%%%%%%%%%%%%%%%%%%%%%%%%%%%%%%%%%%%%%%%%%%%%%%%%%%%%%%%%%%%%%%%%%

    % Decision of the received symbols V polarization:
    DRxBitsV = Decision_DiffEnc(Rx(:,1),ModFormat);

    % Counting the bit error rate:
    BER.V = mean(DRxBitsV ~= TxBits(:,1));

    % If pol. multiplexing is used:
    if NPol == 2
        % Decision of the received symbols in H polarization:
```

```
            DRxBitsH = Decision_DiffEnc(Rx(:,2),ModFormat);

            % Counting the bit error rate:
            BER.H = mean(DRxBitsH ~= TxBits(:,2));
        end
end
```

6.5.4   Auxiliary Functions

Matlab/Octave Code 6.11   BER and SER evaluation for sequential decoding

```
function [BER,SER] = ErrorCounting(Rx,Tx,TxBits,ModFormat,NPol,FPhaseTest)
%%%%%%%%%%%%%%%%%%%%%%%%%%%%%%%%%%%%%%%%%%%%%%%%%%%%%%%%%%%%%%%%%%%%%%%%%%%%%
% ErrorCounting [BER,SER] = ErrorCounting(Rx,Tx,TxBits,ModFormat,NPol,...  %
%                                   FPhaseTest)                            %
%                                                                         %
%    This function evaluates the bit error rate (BER) and the symbol error %
% rate (SER) using the transmitted 'Tx' and received 'Rx' sequences. For  %
% BER evaluation, the received bit sequence is compared to the tranmitted %
% bit sequence 'TxBits'. The received bit sequence is generated after hard%
% decide the received symbols and get their binary labels. For SER        %
% evaluation, symbols of sequence 'Rx' are hard-decided and compared to   %
% the symbols of sequence 'Tx'. Both sequences 'Rx' and 'Tx' must be      %
% normalized to unitary power and synchronized prior to error evaluation. %
% This function considers Gray mapping, with the labels of the function   %
% 'SymbolGeneration'.                                                     %
%                                                                         %
% Input:                                                                  %
%   Rx        = Sequence of received symbols normalized to unitary power*.%
%   Tx        = Sequence of transmitted symbols normalized to unitary     %
%               power*.                                                   %
%   TxBits    = Sequence of transmitted bits associated with sequence     %
%               'Tx'*;                                                    %
%   ModFormat = Modulation format: 'QPSK' or '16-QAM';                    %
%   NPol      = Number of pol. orientations used;                         %
%   FPhaseTest = Flag to enable ('true') or disable ('false') the         %
%               evaluation of BER/SER for 4 different rotations (0 pi/2    %
%               pi 3*pi/2) applied to the entire sequence 'Rx'. This can  %
%               can be helpful for scenarios where 'Rx' may be a version  %
%               of 'Tx' simply rotated by a multiple of pi/2 (due to the  %
%               phase ambiguity of the constellations), e.g., in simul.   %
%               with phase noise and phase recovery;                      %
% *Note: For transmission in single pol. orientation, sequences 'TxBits', %
%        'Tx', and 'Rx' must be column vectors. For transmission with pol.%
%        multiplexing, sequences 'TxBits', 'Tx', and 'Rx' must be matrices%
%        with two column vectors, where each column vector corresponds to %
%        the sequence transmitted in one pol. orientation.                %
%                                                                         %
% Output:                                                                 %
%   BER = Struct with the evaluated BER per pol. orientation. If NPol == 1%
%         the struct BER has only the field BER.V. If NPol == 2, the      %
%         struct BER has the filed BER.V and BER.H;                       %
%   SER = Struct with the evaluated SER per pol. orientation. If NPol == 1%
%         the struct SER has only the field SER.V. If NPol == 2, the      %
%         struct SER has the filed SER.V and SER.H;                       %
%                                                                         %
% This function is part of the book Digital Coherent Optical Systems;     %
% Darli A. A. Mello and Fabio A. Barbosa;                                 %
%%%%%%%%%%%%%%%%%%%%%%%%%%%%%%%%%%%%%%%%%%%%%%%%%%%%%%%%%%%%%%%%%%%%%%%%%%%%%
```

```
    % Test phase:
    if FPhaseTest
        RotAngle = [0 pi/2 pi 3*pi/2];
    else
        RotAngle = 0;
    end

    % Initializing an auxiliary vector:
    BERTempV = NaN(1,length(RotAngle)); SERTempV = NaN(1,length(RotAngle));

    for ii = 1:length(RotAngle)
        % Test phase:
        if FPhaseTest
            RxRot = Rx(:,1)*exp(1i*RotAngle(ii));
        else
            RxRot = Rx(:,1);
        end

        % Decision of the received symbols V polarization:
        DRxBitsV = Decision(RxRot,ModFormat,true);
        DRxSymbV = Decision(RxRot,ModFormat,false);

        % Counting the bit error rate:
        BERTempV(ii) = mean(DRxBitsV ~= TxBits(:,1));
        SERTempV(ii) = mean(DRxSymbV ~= Tx(:,1));
    end

    % Minumum value of BER:
    BER.V = min(BERTempV) ; SER.V = min(SERTempV);

    % If pol. multiplexing is used:
    if NPol == 2
        % Initializing an auxiliary vector:
        BERTempH=NaN(1,length(RotAngle)); SERTempH=NaN(1,length(RotAngle));

        for ii = 1:length(RotAngle)
            % Test phase:
            if FPhaseTest
                RxRot = Rx(:,2)*exp(1i*RotAngle(ii));
            else
                RxRot = Rx(:,2);
            end

            % Decision of the received symbols in H polarization:
            DRxBitsH = Decision(RxRot,ModFormat,true);
            DRxSymbH = Decision(RxRot,ModFormat,false);

            % Counting the bit error rate:
            BERTempH(ii) = mean(DRxBitsH ~= TxBits(:,2));
            SERTempH(ii) = mean(DRxSymbH ~= Tx(:,2));
        end

        % Minumum value of BER:
        BER.H = min(BERTempH) ; SER.H = min(SERTempH);
    end
end
```

Matlab/Octave Code 6.12 Synchronization of transmitted and received sequences using cross-correlation

```
function [TxBits,Tx,Rx] = SyncSignals(TxBits,Tx,Rx,ModFormat,NPol)
%%%%%%%%%%%%%%%%%%%%%%%%%%%%%%%%%%%%%%%%%%%%%%%%%%%%%%%%%%%%%%%%%%%%%%%%%%%%%
% SyncSignals [TxBits,Tx,Rx] = SyncSignals(TxBits,Tx,Rx,ModFormat,NPol)   %
%                                                                         %
%    This function synchronizes the sequence of transmitted symbols 'Tx' to%
% the sequence of received symbols 'Rx'. If needed, symbols of the        %
% sequence 'Tx' are discarded so that, at the output, both sequences have %
% the same length. Note: Symbols of sequence 'Rx' may also be discarded to%
% guarantee that 'Tx' and 'Rx' are the same length at the output. The sequence%
% of transmitted bits associated with the 'synchronized' sequence 'Tx' is %
% also generated.                                                         %
%                                                                         %
% Input:                                                                  %
%   TxBits = Sequence of transmitted bits associated with sequence 'Tx'*; %
%   Tx     = Sequence of transmitted symbols normalized to unitary power*;%
%   Rx     = Sequence of received symbols normalized to unitary power*;   %
%   NPol   = Number of pol. orientations used;                            %
%   *Note  = For transmission in single pol. orientation, 'TxBits', 'Tx', %
%            and 'Rx' must be column vectors. For transmission with pol.  %
%            multiplexing, 'TxBits', 'Tx', and 'Rx' must be matrices with %
%            two column vectors, where each column vector corresponds to  %
%            the signal of one pol. orientation;                          %
%                                                                         %
% Output:                                                                 %
%   Tx,Rx = Synchronized sequences. Sequences 'Tx' and 'Rx' are the same  %
%           length.                                                       %
%   TxBits = Sequence of transmitted bits associated with sequence 'Tx';  %
%                                                                         %
% This function is part of the book Digital Coherent Optical Systems;     %
% Darli A. A. Mello and Fabio A. Barbosa;                                 %
%%%%%%%%%%%%%%%%%%%%%%%%%%%%%%%%%%%%%%%%%%%%%%%%%%%%%%%%%%%%%%%%%%%%%%%%%%%%%

    % Paremeters related to the modulation format:
    switch ModFormat
        case 'QPSK'
            m = 2 ; AbsCorr = false;
        case '16QAM'
            m = 4 ; AbsCorr = true;
    end

    % Auxiliary variables:
    if NPol == 2
        Aux = 4;
    else
        Aux = 1;
    end
    Ratio = zeros(Aux,1) ; Pos   = zeros(Aux,1);

    % Cross-corr. between Tx 1 and Rx 1:
    [Pos(1),Ratio(1)] = Corr(Tx(:,1),Rx(:,1),AbsCorr);

    % If pol. multiplexing is used:
    if NPol == 2
        % Cross-corr. between Tx 2 and Rx 1:
        [Pos(2),Ratio(2)] = Corr(Tx(:,2),Rx(:,1),AbsCorr);

        % Cross-corr. between Tx 2 and Rx 2:
        [Pos(3),Ratio(3)] = Corr(Tx(:,2),Rx(:,2),AbsCorr);

        % Cross-corr. between Tx 1 and Rx 2:
        [Pos(4),Ratio(4)] = Corr(Tx(:,1),Rx(:,2),AbsCorr);
    end

    % Checking if the pol. orientations are switched:
```

```
    if NPol == 2
        AuxFlip = false;
        if Ratio(2) >= Ratio(1) && Ratio(4) >= Ratio(3)
            AuxFlip = true ; Aux     = [1 3];
        elseif Ratio(2) >= Ratio(1) && Ratio(3) >= Ratio(4)
            % Signals of pol. V and H may be switched:
            if Ratio(2) >= Ratio(3)
                AuxFlip = true;
            end
            Aux = [1 3];
        elseif Ratio(1) >= Ratio(2) && Ratio(4) >= Ratio(3)
            % Signals of pol. V and H may be switched:
            if Ratio(4) >= Ratio(1);
                AuxFlip = true;
            end
            Aux = [2 4];
        else
            Aux = [2 4];
        end
        % Updating control variables:
        Ratio(Aux) = [] ; Pos(Aux) = [];
        if AuxFlip
            % Signals of pol. V and H are switched:
            Rx = fliplr(Rx) ; Pos = flipud(Pos) ; Ratio = flipud(Ratio);
        end
    end

    % Defining initial position according to the corr. with largest ratio:
    [~,IPos] = max(Ratio); PosIn = Pos(IPos); PosFin = PosIn+size(Rx,1)-1;
    if PosFin > size(Tx,1)
        % Updating final position:
        PosFin = size(Tx,1);

        % Adjusting sequence Rx so as Tx and Rx are the same length:
        PosInRx = PosIn - size(Tx,1) + size(Rx,1);  Rx = Rx(PosInRx:end,:);
    end

    % Synchronized symbol (Tx) and bit (TxBits) sequences:
    Tx = Tx(PosIn:PosFin,:) ; TxBits = TxBits((PosIn-1)*m+1:PosFin*m,:);
end
```

Matlab/Octave Code 6.13 Cross-correlation calculation

```
function [Pos,Ratio] = Corr(Tx,Rx,AbsCorr)
%%%%%%%%%%%%%%%%%%%%%%%%%%%%%%%%%%%%%%%%%%%%%%%%%%%%%%%%%%%%%%%%%%%%%%%%%%%%%%
% Corr [Pos,Ratio] = Corr(Tx,Rx,AbsCorr)                                   %
%                                                                          %
%    This function calculates cross-correlation between the transmitted    %
% sequence, 'Tx', and the received sequence, 'Rx', and produces 2 outputs. %
% The first one is the position 'Pos' of the maximum value of              %
% cross-correlation between 'Tx' and 'Rx'. The second one is the ratio     %
% between the maximum and the mean value of the cross-correlation, 'Ratio' %
%                                                                          %
% Input:                                                                   %
%   Tx      = Sequence of transmitted symbols normalized to unitary power*.%
%   Rx      = Sequence of received symbols normalized to unitary power*.    %
%   AbsCorr = Flag that determines if the cross-correlation is calculated  %
%             using only the magnitude ('true') or the complex values      %
%             ('false') of the symbols of sequences 'Tx' and 'Rx'. For     %
%             QPSK signals, 'AbsCorr' must be always 'false'. For 16-QAM    %
%             signals, a better synchronization is achieved when 'AbsCorr' %
%             is 'true';                                                    %
%   *Note: Sequences 'Tx' and 'Rx' are column vectors, corresponding to the%
%          signals of one pol. orientation;                                %
%                                                                          %
% Output:                                                                  %
```

```
%   Pos   = Position of the maximum value of the cross-correlation between%
%           'Tx' and 'Rx'.                                                %
%   Ratio = Ratio between the maximum and the mean value of the cross-    %
%           correlation between 'Tx' and 'Rx';                            %
%                                                                         %
% This function is part of the book Digital Coherent Optical Systems;     %
% Darli A. A. Mello and Fabio A. Barbosa;                                 %
%%%%%%%%%%%%%%%%%%%%%%%%%%%%%%%%%%%%%%%%%%%%%%%%%%%%%%%%%%%%%%%%%%%%%%%%%%%%%

    % Cross-correlation calculation:
    if ~AbsCorr
        Corr = xcorr(Tx-mean(Tx),Rx-mean(Rx));
    else
        Corr = xcorr(abs(Tx)-mean(abs(Tx)),abs(Rx)-mean(abs(Rx)));
    end

    % Finding the position of the maximum of the cross-correlation:
    Corr = Corr(size(Tx,1):end) ; [~,Pos] = max(abs(Corr));

    % Ratio between maximum and mean values:
    Ratio = 10*log10(max(abs(Corr))/mean(abs(Corr)));
end
```

References

1. T. Okoshi, Recent advances in coherent optical fiber communication systems. J. Lightwave Technol. **5**(1), 44–52 (1987)
2. E. Ip, J.M. Kahn, Feedforward carrier recovery for coherent optical communications. J. Lightwave Technol. **25**(9), 2675–2692 (2007). [Online]. Available: https://jlt.osa.org/abstract.cfm?URI=jlt-25-9-2675
3. A. Leven, N. Kaneda, U.V. Koc, Y.K. Chen, Frequency estimation in intradyne reception. IEEE Photon. Technol. Lett. **19**(6), 366–368 (2007)
4. M. Selmi, Y. Jaouen, P. Ciblat, Accurate digital frequency offset estimator for coherent polmux QAM transmission systems, in *Proceedings of European Conference on Optical Communication (ECOC)*, pp. 1–2 (2009)
5. T. Pfau, S. Hoffmann, R. Noé, Hardware-efficient coherent digital receiver concept with feedforward carrier recovery for M-QAM constellations. J. Lightwave Technol. **27**(8), 989–999 (2009)
6. S.J. Savory, Digital coherent optical receivers: algorithms and subsystems. IEEE J. Sel. Top. Quantum Electron. **16**(5), 1164–1179 (2010)
7. K.-P. Ho, *Phase-Modulated Optical Communication Systems* (Springer, Boston, MA, 2005)
8. A. Demir, Nonlinear phase noise in optical-fiber-communication systems. J. Lightwave Technol. **25**(8), 2002–2032 (2007)
9. J.P. Gordon, L.F. Mollenauer, Phase noise in photonic communications systems using linear amplifiers. Optics Letters **15**(23), 1351–1353 (1990)
10. M.S. Faruk, S.J. Savory, Digital signal processing for coherent transceivers employing multilevel formats. J. Lightwave Technol. **35**(5), 1125–1141 (2017)
11. M. Morelli, U. Mengali, Feedforward frequency estimation for PSK: a tutorial review. Eur. Trans. Telecommun. **9**(2), 103–116 (1998). [Online]. Available: https://doi.org/10.1002/ett.4460090203
12. J. Salz, Modulation and detection for coherent lightwave communications. IEEE Commun. Mag. **24**(6), 38–49 (1986)
13. A. Viterbi, Nonlinear estimation of PSK-modulated carrier phase with application to burst digital transmission. IEEE Trans. Inf. Theory

14. H.B. Ferreira, Recuperação de portadora para sistemas ópticos com modulação M-PSK e detecção coerente. Bachelor's thesis, University of Brasilia, Brasilia, Brazil, 2009. [Online]. Available: https://bdm.unb.br/handle/10483/1407

15. H.B. Ferreira, Algoritmos de recuperação de fase para sistemas ópticos com modulação DP-QPSK. Master's thesis, University of Brasilia, Brasilia, Brazil, 2012. [Online]. Available: https://repositorio.unb.br/handle/10482/10417

16. E. Alpman, Estimation of oscillator phase noise for MPSK-based communication systems over AWGN channels. Master's thesis, Chalmers University of Technology, Gothenburg, Sweden, 2004

17. V. Rozental, D. Kong, B. Corcoran, D. Mello, A.J. Lowery, Filtered carrier phase estimator for high-order QAM optical systems. J. Lightwave Technol. 36(14), 2980–2993 (2018)

18. T. Koike-Akino, K. Kojima, D. Millar, K. Parsons, Y. Miyata, W. Matsumoto, T. Sugihara, T. Mizuochi, Cycle slip-mitigating turbo demodulation in LDPC-coded coherent optical communications, in *Proceedings of Optical Fiber Communication Conference (OFC)* (Optical Society of America, 2014), p. M3A.3. [Online]. Available: https://www.osapublishing.org/abstract.cfm?URI=OFC-2014-M3A.3

19. H. Cheng, Y. Li, F. Zhang, J. Wu, J. Lu, G. Zhang, J. Xu, J. Lin, Pilot-symbols-aided cycle slip mitigation for DP-16QAM optical communication systems. Optics Express 21(19), 22166–22172 (2013)

Chapter 7
Clock Recovery

Digital communications systems process digitized samples of analog waveforms. These samples must be collected at specific instants of time, and imperfections in this process may significantly impair data transmission. Timing errors can be classified into *sampling phase errors* and *sampling frequency errors*.

Sampling phase errors appear when the receiver clock frequency is accurately synchronized with the transmitter clock frequency, but there is a constant lag with respect to the optimal sampling instants. Let us for example assume a binary phase-shift keying (BPSK) signal transmitted with a perfectly square waveform, and received after filtering as shown in Fig. 7.1a. The ideal sampling instants are those exactly at the middle of the symbol period, collecting most of the symbol energy and mitigating the influence of additive noise. However, if sampling occurs with a lag of a fraction of the symbol period τ, as shown in Fig. 7.1b, the obtained samples have a lower energy, and are more susceptible to noise. The worst-case situation occurs when the sampling instants coincide with the transition points between neighboring symbols, leading to the complete elimination of the transmitted sequence, leaving only noise to be processed. Fortunately, this type of error can be corrected, to some extent, in systems that employ adaptive equalization with fractionally spaced filters, such as current digital coherent optical systems.[1] These filters interpolate the ideal sampling instants, recovering the original transmitted sequence [2].

The consequence of a sampling frequency error is depicted in Fig. 7.1c. In this example, the symbol period T_s is shorter than the sampling period T_a. Although the first symbol is correctly sampled, the frequency sampling error leads to an increasing sampling phase error, resulting in a symbol that is not even sampled. In systems with sampling frequency errors the optimal sampling instants vary over time, leading to a time-varying frequency response that cannot be tracked indefinitely. Therefore,

---

[1]In digital coherent optical systems the analog-to-digital converter (ADC) samples the received signal in sampling rates lower than 2 Sa/Symbol, e.g., 1.5 Sa/Symbol, to save complexity [1]. The signal is usually up-sampled to 2 Sa/Symbol before fractionally spaced equalization.

© Springer Nature Switzerland AG 2021
D. A. de Arruda Mello, F. A. Barbosa, *Digital Coherent Optical Systems*,
Optical Networks, https://doi.org/10.1007/978-3-030-66541-8_7

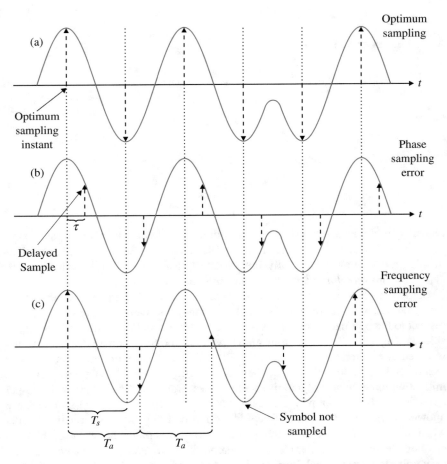

Fig. 7.1 (**a**) Received signal after optimal sampling. (**b**) Phase sampling error. Although transmitter and receiver clocks have the same frequency $1/T_s$, where T_s is the symbol period, there is a lag τ between the optimum and receiver sampling times. (**c**) Frequency sampling errors. Signals are transmitted with a symbol period T_s, but are sampled with a longer sampling period $T_a > T_s$. This mismatch can lead to symbols that are not even sampled. An analogous analysis can be carried out with $T_a < T_s$

clock recovery algorithms, which adaptively correct frequency sampling errors, are of primary importance.

7.1   Clock Recovery Architectures

Clock recovery schemes in digital systems can be classified according to their architecture into three categories [3]. In a purely analog scheme (see Fig. 7.2a), an analog input circuit extracts timing information from the signal and acts directly

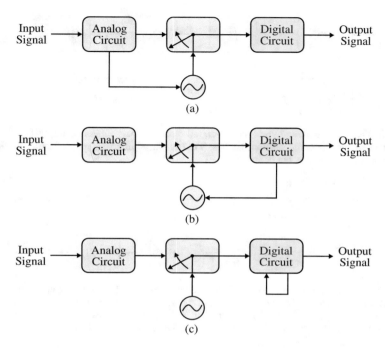

Fig. 7.2 Basic clock recovery schemes [3]. (**a**) Purely analog clock recovery. An analog clock recovery circuit acts directly on the sampling clock. (**b**) Hybrid digital-analog clock recovery. A digital circuit feeds the sampling clock with timing information. (**c**) Purely digital clock recovery structure. Timing error detection and correction is performed digitally. Digital coherent optical systems use combinations of hybrid digital-analog schemes and purely digital schemes

on the sampling clock, synchronizing the input symbol rate with the sampling rate of the analog-to-digital converter (ADC). Such analog clock recovery structures are not used in digital coherent optical receivers, as timing error detection can be effectively implemented in digital domain. In the hybrid scheme (see Fig. 7.2b), the received signal is sampled and processed in the digital domain for extracting timing information. This information is then used to act on the sampling clock. Finally, in the purely digital scheme (see Fig. 7.2c), the extraction of the clock signal, as well as the sampling time adjustments is done in a purely digital way [4]. There are several viable alternatives for implementing clock recovery in digital coherent optical systems, including hybrid analog-digital schemes [5, 6], purely digital schemes [7, 8], or a combination of both [9]. A common solution combines a hybrid analog-digital scheme for coarse timing adjustments, and a purely digital scheme for fine timing adjustments.

Clock recovery algorithms can be implemented in the time or frequency domains. Algorithms in frequency domain can be advantageous from a complexity standpoint if the chain of DSP algorithms already implements fast Fourier transforms (FFTs) and inverse fast Fourier transforms (IFFTs) for other applications, such as chromatic

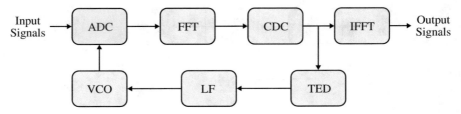

Fig. 7.3 Frequency-domain hybrid analog-digital clock recovery scheme using FFTs and IFFTs implemented for CD compensation [6]. The TED is placed after the multiplication by the CD compensation coefficients, but before the IFFT

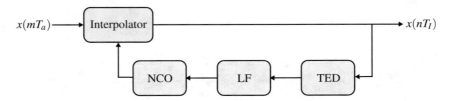

Fig. 7.4 Time-domain purely digital clock recovery scheme. The digital input signal $x[m] = x(mT_a)$ is first interpolated to produce an output signal $x[n] = x(nT_I)$. In order to have the correct sampling frequency, i.e., $T_I \approx T_s/2$ (supposing $T_s/2$-spaced samples), where T_s is the symbol period, a DPLL is required. The DPLL consists of a TED, a LF, and an NCO

dispersion (CD) compensation, avoiding repeated operations. A typical frequency-domain hybrid scheme deployed in optical communications systems is shown in Fig. 7.3 [5, 6]. A timing error detector (TED) signal is obtained in frequency domain after the signal is multiplied by the CD compensation filter, but before the IFFT. The TED output provides a control signal that, after filtering in the loop filter (LF), is able to control the voltage-controlled oscillator (VCO) that actuates on ADCs. A typical time-domain purely digital scheme is shown in Fig. 7.4. The input signal $x(t)$ is first sampled by the ADC at a fixed sampling rate $1/T_a$, where T_a is the sampling period. Thus, a discrete-time digitized signal, with samples $x[m] = x(mT_a)$, is produced at the output of the ADC. The digital signal is then forwarded to an interpolator, which samples the signal at the desired sampling times. The output of the interpolator generates samples with rate $1/T_I$. Ideally, as digital coherent optical systems work with $T_s/2$ fractionally spaced equalizers, $T_I = T_s/2$. The feedback path starts with a TED, providing a control signal that, after the LF, is able to control a numerically controlled oscillator (NCO). The output of the NCO actuates on the interpolator, closing the digital phase-locked loop (DPLL). Several other combinations are also possible, e.g., with purely digital frequency-domain clock recovery [8], but covering all of them would exceed the scope of this book. The next sections detail the main components that are present in most clock recovery schemes.

7.2 Interpolator

In purely digital clock recovery schemes, the purpose of interpolation is to obtain a sample of the signal waveform at a certain instant, based on samples collected at neighboring instants, as depicted in Fig. 7.5. A set of T_a-spaced input samples, indicated at the top of the figure, are used to generate the output sample at instant nT_I. The process is based on interpolating a sample located at a fractional interval $\mu_n T_a$, having as a base point a sample at instant $m_n T_a$, where $m_n = \left\lfloor \frac{kT_I}{T_a} \right\rfloor$. Interpolation has already been discussed in Chap. 4, in the scope of skew compensation. Clock recovery uses similar algorithms, which are based on transversal linear filters. However, unlike skew compensation, the clock recovery scheme adjusts the filter coefficients dynamically, on a symbol-by-symbol basis, based on the output of an NCO. Interpolation can be implemented in frequency domain along with CD compensation [8], or using a time-domain interpolator. A popular solution in time domain uses a cubic Lagrange interpolator (for a detailed explanation of Lagrange interpolation, please see Chap. 4). Cubic interpolators can have 4–10 taps with increasing passband [1]. The coefficients of a 4-tap cubic interpolator, for an interpolation instant $t = m_n T_a + \mu_n T_a$, are given by Erup et al. [10]

$$
\begin{aligned}
w_{n-2} &= -\frac{1}{6}\mu_n^3 + \frac{1}{6}\mu_n; \\
w_{n-1} &= \frac{1}{2}\mu_n^3 + \frac{1}{2}\mu_n^2 - \mu_n; \\
w_n &= -\frac{1}{2}\mu_n^3 - \mu_n^2 + \frac{1}{2}\mu_n + 1; \\
w_{n+1} &= \frac{1}{6}\mu_n^3 + \frac{1}{2}\mu_n^2 + \frac{1}{3}\mu_n.
\end{aligned}
\tag{7.1}
$$

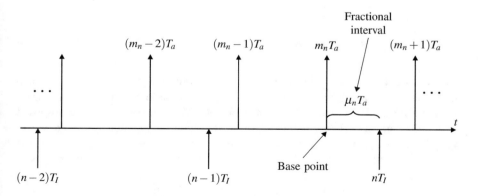

Fig. 7.5 Interpolation for clock recovery. T_a-spaced input samples are used to generate T_I-spaced output samples. In the figure, $m_n T_a$ is the base point and $\mu_n T_a$ is the fractional interval

In

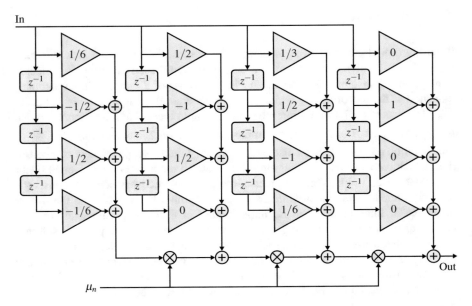

Fig. 7.6 Cubic interpolator with Farrow architecture. The filter coefficients remain fixed, and the fractional interval μ_n is updated in every clock cycle [10, 11]

Updating the filter coefficients for every fractional interval μ_n would require the calculation and transfer of four coefficients. Alternatively, the interpolation structure proposed by Farrow in [11] requires the transfer of a single variable μ_n, at the cost a more complex filter structure, as shown in Fig. 7.6. In the Farrow architecture, the lines of multipliers correspond to the terms involved in the calculation of w_{n+1} (upper line) to w_{n-2} (bottom line), and the columns correspond to multiplications by μ_n^0 (right column) to μ_n^3 (left column).

7.3   Timing Error Detector

The TED extracts timing information and forwards it, after filtering, to a VCO or NCO. This process in carried out in closed-loop, building the so-called phase-locked loop (PLL). Perhaps the most deployed TED scheme is the one proposed by Gardner in [12]. Taking three $T_s/2$-spaced neighboring samples of the received signal $x[2k-2]$, $x[2k-1]$, and $x[2k]$, assumed real for the sake of simplicity, the Gardner algorithm calculates the T_s-spaced timing error indication signal $e[k]$ as

$$e[k] = x[n-1](x[n] - x[n-2]), \quad n = 2k. \tag{7.2}$$

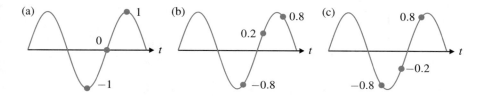

Fig. 7.7 Gardner TED. Three cases are indicated: (**a**) $\Delta\tau = 0$ (correct sampling time), (**b**) $\Delta\tau > 0$ (late sampling), and (**c**) $\Delta\tau < 0$ (early sampling)

The basic principle of the Gardner TED is depicted in Fig. 7.7 for a BPSK signal. The three red circles indicate the input samples used by the algorithm. Three cases are indicated: (a) $\Delta\tau = 0$ (correct sampling time), (b) $\Delta\tau > 0$ (late sampling), and (c) $\Delta\tau < 0$ (early sampling) [13]. In each of these cases we have, respectively,

$$e[k] = 0 \cdot (1 - (-1)) = 0; \tag{7.3}$$

$$e[k] = 0.2 \cdot (0.8 - (-0.8)) = 0.32 > 0; \tag{7.4}$$

$$e[k] = -0.2 \cdot (0.8 - (-0.8)) = -0.32 < 0. \tag{7.5}$$

If the input signal is sampled at the right phase, the central sample coincides with transitions, and the error signal goes ideally to zero. If the sampling phase is late, the central sample has a positive value, and the sign of $e[k]$ is positive. If, however, the sampling phase is too early, the central sample has a negative value, and the sign of $e[k]$ is negative. Therefore, the absolute value of $e[k]$ gives an indication of the magnitude of the sampling phase mismatch, while its sign indicates if the sampling phase is early or late. Although at first sight the Gardner algorithm seems to correct for phase sampling errors only, its continuous operation and the adjustments made over time provide a substantial correction to frequency sampling errors.

In systems with in-phase and quadrature modulation, the TED can combine both components as [7]

$$e[k] = x^I[n-1]\left(x^I[n] - x^I[n-2]\right) + x^Q[n-1]\left(x^Q[n] - x^Q[n-2]\right)$$
$$= \Re\left\{x^*[n-1]\left(x[n] - x[n-2]\right)\right\}, \quad n = 2k. \tag{7.6}$$

A TED is usually characterized by its so-called S-curve, which is obtained by sweeping the phase mismatch $\Delta\tau$ in the interval $[-T_s/2, T_s/2]$ and observing its response. Figure 7.8 shows the S-curve for the Gardner TED. As discussed earlier, deviations in the correct sampling instant $\Delta\tau$ cause an increase in the magnitude of output $e[k]$, and its sign indicates a late or early sampling time.

The Gardner TED requires well-defined transitions between adjacent symbols, but optical digital coherent systems are strongly affected by dispersive effects. Figure 7.8 shows the S-curve when the signal is affected by CD and polarization

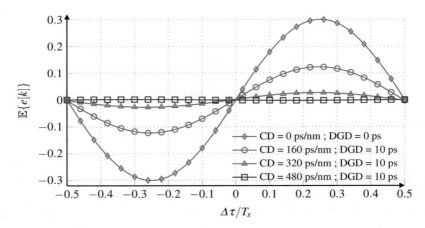

Fig. 7.8 S-curve for the Gardner TED, for a system affected by CD and PMD. Deviations in sampling phase $\Delta\tau$ ($[-T_s/2, T_s/2]$) cause an increase in the magnitude of output $e[k]$. Its sign indicates a late or early sampling time. Intersymbol interference caused by CD or PMD severely reduces the S-curve derivative at $\Delta\tau = 0$, affecting the clock recovery performance

mode dispersion (PMD). The derivative of the S-curve near $\Delta\tau = 0$ fades rapidly with the increase of dispersion, severely impairing the timing recovery performance. Time-domain TED algorithms are usually applied after the CD compensation filter, solving half of the problem. The question then is whether it is more appropriate to implement it before or after the butterfly adaptive PMD compensation filter. Implementing it before PMD compensation, as shown in Fig. 7.9a, preserves the equalizer from stochastic changes due to clock variations. On the other hand, some pathological cases of PMD can make the clock signal disappear, deteriorating the system performance. A possible solution to this problem is to undo the polarization rotation carried out by the channel combining the signals of the V and H orientations, but this strategy requires a tracking loop for the rotation angle [14]. Furthermore, the polarization-related channel effects can be more complicated than a single differential group delay (DGD) element.

While Gardner's TED performs satisfactorily with NRZ and RZ pulses, it is strongly impaired by Nyquist pulses with small roll-off factors. The algorithm proposed in [15] implements a modified Gardner (MG) TED based on the power of the equalized symbols, as indicated in Fig. 7.9b.[2] The MG TED output is given by

$$e[k] = y[n - 1]y[n - 1]^* \left(y[n - 2]y[n - 2]^* - y[n]y[n]^*\right), \quad n = 2k, \qquad (7.7)$$

---

[2] A power-based TED strategy is also used in [16], estimating timing errors before adaptive equalization. In this case, solutions to track polarization issues may be required.

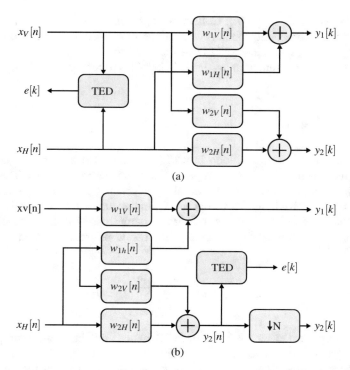

Fig. 7.9 (a) Time-domain TED placed before the adaptive equalizer. (b) Time-domain modified Gardner (MG) TED for Nyquist-shaped signals with low roll-off factors. The TED is placed after the adaptive equalizer. Filters w_{2V} and w_{2H} generate $T_s/2$-spaced samples, whereas filters w_{1V} and w_{1H} generate T_s-spaced samples

where $y[n]$ is the $T_s/2$-spaced output of the equalizer. The MG TED requires $T_s/2$-spaced samples after the adaptive equalizer. As the adaptive equalizer usually outputs T_s-spaced samples, the equalizer complexity is higher than usual. Therefore, the MG TED is applied to only one of the outputs, leaving the other operating with the original structure and generating T_s-spaced samples. The S-curve for the MG TED becomes more pronounced with reduced roll-off factors, while the standard Gardner TED exhibits the opposite behavior. Therefore, the MG TED should be applied only with low roll-off factors.

So far we have discussed TED implementations in the time domain. Implementing the TED in frequency domain can save complexity if the chain of DSP algorithms already use FFTs/IFFTs for other operations, e.g., for CD compensation. An effective solution is the Godard TED, given by Godard [17]

$$e[k] = \sum_{n=0}^{N/2-1} \text{Im}\{X[n] \cdot X^*[n + N/2])\}, \tag{7.8}$$

where $X[n]$ is the FFT of $x[n]$. Collecting $X[n]$ inside the CD equalization filter saves additional complexity, but an eventual loss of clock tone because of PMD can still arise. Although the Godard TED was devised for $T_s/2$-spaced samples, it can also be adapted to operate with lower sampling rates to reduce complexity [8].

7.4   Loop Filter

The stability of the feedback loop in a PLL is controlled by the LF.[3] Usually, a proportional-integral (PI) filter is applied [7], as shown in Fig. 7.10. The proportional arm multiplies the received signal by constant k_p, while the integral arm multiplies the received signal by constant k_i and accumulates all current and past values. The integral arm ensures a constant output case the input reaches zero (indicating perfect timing). The outputs of the two arms are then added to generate output $W[k]$

$$W[k] = e[k]k_p + k_i \sum_{l=0}^{k-1} e[k-l]. \tag{7.9}$$

The system created by the VCO or NCO (whose output phase is proportional to the integral of the control signal), and the PI LF (which also has an integrator), forms a second-order PLL. The choice of k_p and k_i is critical for the proper PLL operation [19], and is highly dependent on the system peculiarities. Some directions for selecting k_p and k_i for typical optical communications systems are provided in [7], based on the parameters of an analog PLL. There are two design parameters

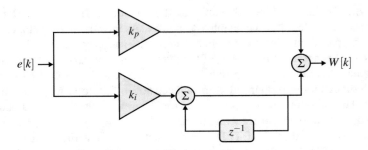

Fig. 7.10 Proportional-integral filter used for the PLL. The proportional coefficient is given by k_p, while the integral coefficient by k_i. The integral arm with the z^{-1} loop implements an accumulator. Output $W[k]$ is used to control the NCO

---

[3] Although it is called a *loop filter* in the theory of PLLs, a better name might have been a *loop controller*, as its main purpose is to control the feedback loop [18, p. 11].

involved in the calculations of k_p and k_i for an analog PLL, namely the noise bandwidth B_L and the damping ratio ζ. The noise bandwidth B_L is a parameter which governs the trade-off between the ability to track fast clock fluctuations and the filtering of noise. Therefore, oscillators with high clock excursions require higher values of B_L. The damping ratio ζ is usually selected between 0.5 and 2, being 0.7 a typical choice. The proportional and integral gains are given by Gardner [18, Eq. 2.16]

$$k_i = \frac{\omega_n^2}{k_d}, \tag{7.10}$$

$$k_p = \frac{2\zeta}{\sqrt{k_d / k_i}}, \tag{7.11}$$

where k_d is the TED sensitivity, given by the derivative of the S-curve around $\Delta t = 0$. The natural frequency of the feedback loop, ω_n, is given by Gardner [18, p. 130]

$$\omega_n = \frac{8 B_L \zeta}{1 + 4\zeta^2}. \tag{7.12}$$

In typical purely digital clock recovery, B_L can vary between 0.5×10^{-3} and 5×10^{-3} depending on the maximal clock offset to be corrected [7].

7.5 Numerically Controlled Oscillator

In purely digital clock recovery, the filtered control signal provided by the LF is sent to the NCO, which produces the fractional interval and underflow signals that feed the interpolator [3], completing the DPLL. The oscillating signal produced by the NCO is the output of a mod-1 operator

$$\eta_{m_n}^{NCO} = \left[\eta_{m_n-1}^{NCO} - W[k] \right]_{\text{mod-1}}, \tag{7.13}$$

where $W[k]$ is the LF output at the time of input sample m_n. The mod-1 operator is equal to the remainder of a division by 1. In the division, the quotient can be zero, positive, and negative, while the remainder is always in interval $[0, 1)$. For example, if the argument is 0.8, the modulus operation yields 0.8 (the quotient is 0 and the remainder is 0.8). If the argument is -0.8, the modulus operation yields 0.2 (the quotient is -1 and the remainder is 0.2). When the interpolation interval T_I reaches exactly the symbol period T_s, the TED output becomes 0. Consequently, the LF proportional arm becomes 0 and the integral arm stops accumulating, generating a constant output $W[k]$. Under these conditions, the NCO decreases by $W[k]$ every T_a seconds (according to (7.13)), causing the NCO to underflow in average after $1/W[k]$ clock ticks [3]. The NCO period T_I is exactly the period between underflows. Therefore, it is given by the product of the clock period T_a, and the average number of clock ticks between underflows $1/W[k]$, i.e., $T_I \approx T_a/W[k]$. Therefore, $W[k]$ in steady-state is given by Gardner [3]

$$W[k] \approx \frac{T_a}{T_I}. \tag{7.14}$$

In steady-state, the output of the LF approximates the ratio between T_a and T_I. The main issue now is how to obtain the fractional interval and the base point to feed the interpolator. The fractional interval μ_n for sample $x(nT_I)$ can be obtained through the relationships observed in Fig. 7.11. At instant $m_n T_a$ the NCO output is $\eta_{m_n}^{\mathrm{NCO}}$. As the NCO decays by $W[k]$ every T_a seconds, its output before the mod-1 operation after T_a seconds is $\eta_{m_n}^{\mathrm{NCO}} - W[k]$. Knowing that the NCO output crosses zero exactly at $nT_I = (m_n + \mu_n)T_a$, and with simple algebraic manipulations, we can obtain the following relationship [3]:

$$\mu_n \approx \frac{\eta_{m_n}^{\mathrm{NCO}}}{W[k]}. \tag{7.15}$$

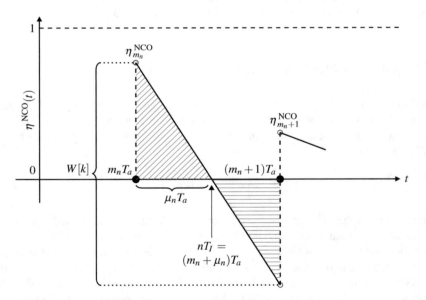

Fig. 7.11 NCO operation and derivation of the fractional interval μ_n. At instant $m_n T_a$ the NCO output is $\eta_{m_n}^{\mathrm{NCO}}$. As the NCO decays by $W[k]$ every interval T_a, its output before the mod-1 operation at instant $(m_n + 1)T_a$ is $\eta_{m_n}^{\mathrm{NCO}} - W[k]$. Knowing that the NCO output crosses zero exactly at $nT_I = (m_n + \mu_n)T_a$, simple algebraic manipulations yield $\mu_n \approx \eta_{m_n}^{\mathrm{NCO}}/W[k]$

The base point can be identified from the following rules [20]:

1. If $-1 \leq \eta_{m_n}^{NCO} - W[k] < 0$, then $m_{n+1} = m_n + 1$. In this case the next sample m_{n+1} is used as base point for the next interpolator update.
2. If $\eta_{m_n}^{NCO} - W[k] \geq 0$, then $m_{n+1} = m_n + 2$. In this case one sample is skipped, and the other sample is used as base point for the next interpolator update.
3. If $\eta_{m_n}^{NCO} - W[k] < -1$, $m_{n+1} = m_n$. In this case two consecutive interpolator updates use the same base point m_n.

In order to illustrate the NCO operation, let us consider two simple examples. In the first one, shown in Fig. 7.12, $T_I = 1.1 T_a$ and, according to (7.14), $W[k] \approx 0.9$. The blue dots indicate outputs $\eta_{m_n}^{NCO}$ of the NCO. The T_I-spaced samples are obtained from the crossings points of the diagonal lines and the horizontal time axis. As the index n increases, the fractional interval μ_n and the base point m_n are also increased accordingly. This behavior is observed until $n = 10$, where one sample is skipped as base point because of rule (2). In the second example, shown in Fig. 7.13, $T_I = 0.9 T_a$ and $W[k] \approx 1.1$. As the index n increases, the fractional interval μ_n decreases and the base point m_n increases. This behavior is observed until $n = 10$, where sample 10 is repeated as base point because of rule (3).

7.6 Problems

1. Generate a Nyquist-shaped 16-QAM signal. Use a root-raised cosine (RRC) filter with roll-off factor $\beta^{RC} = 0.5$ and a span of 20 symbols, assuming 16 Sa/Symbol. Plot the S-Curve for the Gardner TED. To do this, filter the signal with a matched filter and downsample the generated sequence to 2 Sa/Symbol for different fractional intervals $\Delta T_s / T_s = \{-0.5, -0.45, \cdots, 0.5\}$. Apply the Gardner TED. Ensure that the signals after sampling have unitary power. Plot the average TED response considering the complete sequence. Repeat the process for $\beta^{RC} = 0.1$. Discuss the results.
2. Repeat Problem 1 using the MG TED instead of the Gardner TED. Discuss the results.
3. Plot the numerical derivative of the S-Curve for the Gardner TED, using RRC pulses with $\beta^{RC} = 0.5$. Estimate parameter k_d, i.e., the derivative at $\Delta\tau = 0$.
4. Design the parameters k_p and k_i of an analog proportional-integral loop filter with damping ratio $\zeta = 0.707$. Consider the noise bandwidth values $B_L = 0.5 \times 10^{-3}$ and $B_L = 5 \times 10^{-3}$. Use k_d calculated in Problem 3.
5. Repeat Problem 3 for the MG TED and RRC pulses with $\beta^{RC} = 0.1$.
6. Design parameters k_p and k_i of an analog proportional-integral loop filter with damping ratio $\zeta = 0.707$. Consider the parameter k_d calculated in Problem 5. Evaluate noise bandwidth values $B_L = 0.5 \times 10^{-3}$ and $B_L = 5 \times 10^{-3}$.
7. Simulate the transmission of a single-polarization signal. Assume a Nyquist-shaped 16-QAM signal at 50 GBaud and at OSNR = 25 dB. Use 16-Sa/Symbol

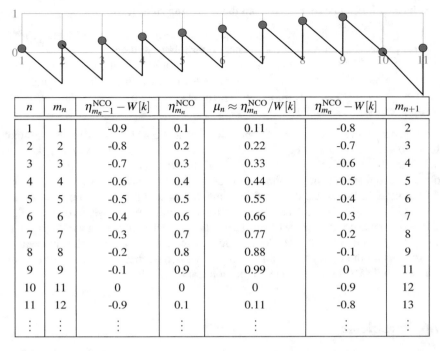

n	m_n	$\eta^{NCO}_{m_n-1} - W[k]$	$\eta^{NCO}_{m_n}$	$\mu_n \approx \eta^{NCO}_{m_n}/W[k]$	$\eta^{NCO}_{m_n} - W[k]$	m_{n+1}
1	1	-0.9	0.1	0.11	-0.8	2
2	2	-0.8	0.2	0.22	-0.7	3
3	3	-0.7	0.3	0.33	-0.6	4
4	4	-0.6	0.4	0.44	-0.5	5
5	5	-0.5	0.5	0.55	-0.4	6
6	6	-0.4	0.6	0.66	-0.3	7
7	7	-0.3	0.7	0.77	-0.2	8
8	8	-0.2	0.8	0.88	-0.1	9
9	9	-0.1	0.9	0.99	0	11
10	11	0	0	0	-0.9	12
11	12	-0.9	0.1	0.11	-0.8	13
⋮	⋮	⋮	⋮	⋮	⋮	⋮

Fig. 7.12 NCO operation with $T_I = 1.1T_a$ and, therefore, $W[k] \approx 0.9$. The blue dots indicate the NCO output η_{m_n}. The crossing points of the diagonal lines with the horizontal time axis indicate the T_I-spaced sampling points. The time intervals between T_I-spaced sampling points and the preceding T_a-spaced sampling points are the fractional intervals μ_n. Note that sample 10 is skipped as base point m_n

RRC pulse shaping with $\beta^{RC} = 0.1$ and a span of 20 symbols.[4] Consider optical modulation and coherent detection in the simulation model. At the receiver, apply a matched filter. Downsample the signal to 2.001 Sa/Symbol, i.e., considering a sampling frequency error of 500 ppm. Implement a DPLL with the MG TED and use it to correct the sampling frequency error on the signal. Use the two sets of parameters calculated in Problem 6. Plot the last 1000 samples of the signal at the output of the DPLL, downsampled to 1 Sa/Symbol. Choose the best set of samples, i.e., even or odd samples. Plot $W[k]$ and observe its convergence. What is the expected steady-state value of $W[k]$? Compare the results obtained with the two sets of parameters, analyzing the constellations and the convergence of $W[k]$.

8. Considering the simulation scenario of Problem 7, evaluate the BER for OSNRs in the interval from 10 dB to 20 dB, and compare them to theoretical values. Use

---

[4]Transmit signals with at least 2^{16} symbols.

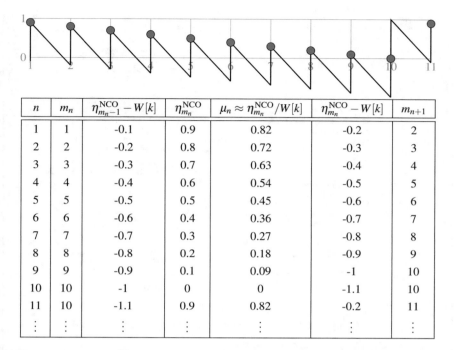

Fig. 7.13 NCO operation with $T_I = 0.9T_a$ and, therefore, $W[k] \approx 1.1$. The blue dots indicate the NCO output at $\eta_{m_n}^{\text{NCO}}$. The crossing points of the diagonal lines with the horizontal time axis indicate the T_I-spaced sampling points. The time intervals between T_I-spaced sampling points and the preceding T_a-spaced sampling points are the fractional intervals μ_n. Note that sample 10 is used twice as a base point

n	m_n	$\eta_{m_n-1}^{\text{NCO}} - W[k]$	$\eta_{m_n}^{\text{NCO}}$	$\mu_n \approx \eta_{m_n}^{\text{NCO}}/W[k]$	$\eta_{m_n}^{\text{NCO}} - W[k]$	m_{n+1}
1	1	-0.1	0.9	0.82	-0.2	2
2	2	-0.2	0.8	0.72	-0.3	3
3	3	-0.3	0.7	0.63	-0.4	4
4	4	-0.4	0.6	0.54	-0.5	5
5	5	-0.5	0.5	0.45	-0.6	6
6	6	-0.6	0.4	0.36	-0.7	7
7	7	-0.7	0.3	0.27	-0.8	8
8	8	-0.8	0.2	0.18	-0.9	9
9	9	-0.9	0.1	0.09	-1	10
10	10	-1	0	0	-1.1	10
11	10	-1.1	0.9	0.82	-0.2	11
⋮	⋮	⋮	⋮	⋮	⋮	⋮

parameters k_p and k_i calculated for $B_L = 0.5 \times 10^{-3}$. Note that a set of samples at the output of the DPLL should be discarded to account for its convergence period.

7.7 Matlab/Octave Functions

7.7.1 Functions for Sect. 7.1

Matlab/Octave Code 7.1 Clock recovery scheme based on the Gardner and the MG TED

```
function [Out,varargout] = ClockRecovery(In,PSType,NSymb,ParamCR)
%%%%%%%%%%%%%%%%%%%%%%%%%%%%%%%%%%%%%%%%%%%%%%%%%%%%%%%%%%%%%%%%%%%%%%%%%%%%%%
% [Out,varargout] = ClockRecovery(In,PSType,NSymb,ParamCR)               %
%                                                                        %
%  This function performs clock recovery in signal 'In' using a DPLL     %
% structure consisting on an interpolator, a TED, a loop filter, and a NCO%
% Signal 'In' must be obtained at 2 Sa/Symbol. The function uses the     &
% Gardner TED for NRZ pulses, and the MG TED for Nyquist shaped pulses.  %
```

```
% The length of the output sequence is limited to the number of          %
% transmitted symbols 'NSymb' times the number of samples per symbol at   %
% the input signal 'In' (2 Sa/Symbol).                                    &
% *Notes:-The clock recovery is not done in the first two samples of 'In'.%
%         -This function is designed to be used before adaptive equalization%
%          However, in the presence of PMD, extra processing for tracking %
%          polarization rotations is required;                            %
%                                                                         %
% Input:                                                                  %
%   In      = Signal obtained at 2 Sa/Symbol of one pol. orientation in   %
%             which clock recovery will be performed. 'In' must be a      %
%             column vector;                                              %
%   PSType  = Type of pulse shaping filter: 'NRZ' or 'Nyquist';           %
%   ParamCR = Struct that specifies parameters of the clock recovery:     %
%           - ParamCR.ki: Constant of the integral part of the loop filter; %
%           - ParamCR.kp: Constant of the proportional part of the loop filter;%
%           - ParamCR.DPLLVarOut: Flag to enable (true) or disable (false) %
%                internal variables of the DPLL (i.e., the output of the loop %
%                filter 'Wk', the NCO output 'Etamn', the fractional interval %
%                'mun', the base point 'mn', and TED output 'ek') as output of%
%                the function;                                            %
%                                                                         %
% Output:                                                                 %
%   xOut     = Signal obtained after clock recovery (column vector);      %
%   varargout = For 'ParamCR.DPLLVarOut = true', internal variables of the%
%                DPLL are also outputs of the function: 'Wk', 'Etamn',    %
%                'mun','mn', and 'ek';                                    %
%                                                                         %
% This function is part of the book Digital Coherent Optical Systems;     %
% Darli A. A. Mello and Fabio A. Barbosa;                                 %
%%%%%%%%%%%%%%%%%%%%%%%%%%%%%%%%%%%%%%%%%%%%%%%%%%%%%%%%%%%%%%%%%%%%%%%%%%%%%

    % Integral and proportional constants:
    ki = ParamCR.ki ; kp = ParamCR.kp;

    % Initializing variables:
    Etamn = 0.5 ; Wk = 1 ; LF_I = Wk ; mun = 0 ; n  = 3 ; mn = n;
    Out = zeros(1,length(In)) ; Out(1:3) = In(1:3) ; LIn = length(In);
    if ParamCR.DPLLVarOut
        WkVec = [] ; EtamnVec = [] ; munVec = []; mnVec = [] ; ekVec = [] ;
    end

    % Clock recovery:
    while mn <= LIn
        if mn == LIn;
            In(mn+1) = 0;
        end

        % Cubic interpolator with Farrow architecture:
        Out(n) = In(mn-2)*(-1/6*mun^3 +    0*mun^2 + 1/6*mun + 0)+ ...
                 In(mn-1)*( 1/2*mun^3 + 1/2*mun^2 -   1*mun + 0)+ ...
                 In(mn)  *(-1/2*mun^3 -   1*mun^2 + 1/2*mun + 1)+ ...
                 In(mn+1)*( 1/6*mun^3 + 1/2*mun^2 + 1/3*mun + 0);

        % Generating the Ts-spaced timing error indication signal 'e_n':
        if mod(n,2)==1
            switch PSType
                case 'Nyquist'
                    % TED - Nyquist pulses:
                    ek = abs(Out(n-1)).^2.*...
                         (abs(Out(n-2)).^2 - abs(Out(n)).^2);
                case 'NRZ'
                    % TED - NRZ pulses:
                    ek = real(conj(Out(n-1)).*(Out(n) - Out(n-2)));
            end

            % Loop Filter:
```

```
            LF_I = ki*ek + LF_I; LF_P = kp*ek; Wk = LF_P + LF_I;

            % Parameters of the DPLL:
            if ParamCR.DPLLVarOut
                WkVec    = [WkVec Wk]    ; EtamnVec = [EtamnVec Etamn];
                munVec   = [munVec mun]  ; mnVec    = [mnVec mn];
                ekVec    = [ekVec ek];
            end
        end

        % NCO - Base point 'mk' and fractional interval 'mu_n':
        if -1 < (Etamn - Wk) && (Etamn - Wk) < 0
            mn = mn + 1;
        elseif (Etamn - Wk) >= C
            mn = mn + 2;
        end
        Etamn = mod(Etamn - Wk,1) ; mun   = Etamn/Wk;

        % Updating the temporal index 'n':
        n = n + 1;
    end

    % Limiting the length of the output to NSymb*2:
    if NSymb*2 < length(Out)
        Out = Out(1:NSymb*2).';
    else
        Out = Out.';
    end

    % Parameters of the DPLL:
    if ParamCR.DPLLVarOut
        varargout{1} = WkVec.'   ; varargout{2} = EtamnVec.';
        varargout{3} = munVec.'  ; varargout{4} = mnVec.'   ;
        varargout{5} = ekVec.'   ;
    end
end
```

References

1. S.M. Bilal, C.R.S. Fludger, Interpolators for digital coherent receivers, in *Proceedings of Photonic Networks; 19th ITG-Symposium* (2018), pp. 1–3
2. G. Ungerboeck, Fractional tap-spacing equalizer and consequences for clock recovery in data modems. IEEE Trans. Commun. **24**(8), 856–864 (1976)
3. F.M. Gardner, Interpolation in digital modems. I. Fundamentals. IEEE Trans. Commun. **41**(3), 501–507 (1993)
4. J. Bergmans, *Digital baseband transmission and recording.* (Kluwer Academic Publishers, Netherlands, 1996)
5. H. Sun, K.-T. Wu, Clock recovery and jitter sources in coherent transmission systems, in *Proceedings of Optical Fiber Communication Conference (OFC)* (Optical Society of America, America, 2012), p. OTh4C.1. http://www.osapublishing.org/abstract.cfm?URI=OFC-2012-OTh4C.1
6. X. Zhou, Efficient clock and carrier recovery algorithms for single-carrier coherent optical systems: a systematic review on challenges and recent progress. IEEE Signal Process Mag. **31**(2), 35–45 (2014)
7. X. Zhou, X. Chen, W. Zhou, Y. Fan, H. Zhu, Z. Li, All-digital timing recovery and adaptive equalization for 112-Gbit/s POLMUX-NRZ-DQPSK optical coherent receivers. IEEE/OSA J. Opt. Commun. Networking **2**(11), 984–990 (2010)

8. A. Josten, B. Baeuerle, E. Dornbierer, J. Boesser, D. Hillerkuss, J. Leuthold, Modified Godard timing recovery for non-integer oversampling receivers. Appl. Sci. **7**(7), 655 (2017)
9. N. Stojanović, X. Chuan, Clock recovery in coherent optical receivers, in *Proceedings of Optical Fiber Communication Conference (OFC)* (2015), pp. 1–3
10. L. Erup, F.M. Gardner, R.A. Harris, Interpolation in digital modems II. Implementation and performance. IEEE Trans. Commun. **41**(6), 998–1008 (1993)
11. C. Farrow, A continuously variable digital delay element, in *IEEE International Symposium on Circuits and Systems*, vol. 3 (1988), pp. 2641–2645
12. F. Gardner, A BPSK/QPSK timing-error detector for sampled receivers. IEEE Trans. Commun. **34**(5), 423–429 (1986)
13. V. Rozental, Hitless rate and bandwidth switching in dynamically reconfigurable coherent optical systems. Ph.D. dissertation (University of Brasilia, Brasilia, 2016). https://repositorio.unb.br/handle/10482/21966
14. H. Sun, K.-T. Wu, A novel dispersion and PMD tolerant clock phase detector for coherent transmission systems, in *Proceedings of the 2011 Optical Fiber Communication Conference and Exposition and the National Fiber Optic Engineers Conference* (2011), pp. 1–3
15. N. Stojanovic, C. Xie, Y. Zhao, B. Mao, N. Gonzalez, J. Qi, N. Binh, Modified Gardner phase detector for Nyquist coherent optical transmission systems, in *Proceedings of Optical Fiber Communication Conference and National Fiber Optic Engineers Conference (OFC/NFOEC)* (Optical Society of America, America, 2013), p. JTh2A.50. http://www.osapublishing.org/abstract.cfm?URI=OFC-2013-JTh2A.50
16. M. Yan, Z. Tao, L. Dou, L. Li, Y. Zhao, T. Hoshida, J.C. Rasmussen, Digital clock recovery algorithm for Nyquist signal, in *Optical Fiber Communication Conference and National Fiber Optic Engineers Conference* (Optical Society of America, America, 2013), p. OTu2I.7. http://www.osapublishing.org/abstract.cfm?URI=OFC-2013-OTu2I.7
17. D. Godard, Passband timing recovery in an all-digital modem receiver. IEEE Trans. Commun. **26**(5), 517–523 (1978)
18. F.M. Gardner, *Phaselock Techniques*, vol. 3 (Wiley, New Jersey, 20050
19. V. Kratyuk, P.K. Hanumolu, U. Moon, K. Mayaram, A design procedure for all-digital phase-locked loops based on a charge-pump phase-locked-loop analogy. IEEE Trans. Circuits Syst. II Express Briefs **54**(3), 247–251 (2007)
20. T.F. Portela, Técnicas de Recuperação de Relógio para Sistemas DP-QPSK. Master's thesis (University of Brasilia, Brasilia, 2012). https://repositorio.unb.br/handle/10482/13345

Chapter 8
Performance Evaluation

8.1 Introduction

The ultimate objective of any digital communication system is to connect applications with tolerable latency and low bit error rate (BER), yielding minimum sustained data rates. Latency depends fundamentally on the link length and on the processing of information at the transmitter and the receiver. The BER depends on several design options, many of which were discussed in previous chapters. We discussed the process of modulation, in which bits of information are mapped into symbols of a certain constellation for transmission over a physical channel. Each symbol of the constellation corresponded to a specific configuration of amplitude and phase of the optical carrier. At the receiver, we detected the modulated optical carrier and attempted to recover back the transmitted bits. One issue that we have not discussed so far is that modern digital coherent optical systems use error control strategies performed by forward error correction (FEC) to achieve the lowest possible BER given certain channel constraints.

In optical communications systems, BERs in the order of 10^{-15} are required after FEC to comply with high-quality demanding services. In previous chapters we computationally evaluated the BER of optical systems by Monte-Carlo simulations, in which a sequence of bits is generated, transmitted over the channel and received. We estimated the BER as the ratio between the number of bit errors and the number of simulated bits. By the law of large numbers, the larger the number of simulated bits, the higher the accuracy of the BER estimate. In Monte-Carlo simulations, usually at least 100 errors are required to achieve a minimally reliable measure of BER. Therefore, in order to simulate a BER of 10^{-15}, in average 10^{17} bits should be generated. Generating such high number of bits is time-consuming and sometimes impractical with the available computational resources. To avoid this problem, performance evaluation in optical communication systems has typically relied on approaches that estimate the pre-FEC BER (BER_{pre}), and then infer the post-FEC BER (BER_{post}) considering the expected FEC performance.

© Springer Nature Switzerland AG 2021
D. A. de Arruda Mello, F. A. Barbosa, *Digital Coherent Optical Systems*,
Optical Networks, https://doi.org/10.1007/978-3-030-66541-8_8

The first generations of FEC schemes deployed in optical communications used hard-decision (HD) decoding, in which the FEC decoder operates on a sequence of symbols or bits after decision. Assuming that symbol errors are uncorrelated, HD decoding allows for a one-to-one mapping between BER_{pre} and BER_{post}. Even if symbol errors are correlated, an interleaver can decorrelate eventual error bursts [1]. Therefore, the BER_{pre} has been used for many years as the reference metric for performance evaluation in optical communications. The system was designed for a certain BER_{pre}, aiming at a certain BER_{post} at the output of the FEC decoder. However, more recent generations of FEC schemes used in optical communications are based on soft decisions. Soft-decision (SD) decoders exploit reliability information usually in the form of log-likelihood ratios (LLRs), instead of (hard) decisions made on the received symbols.[1] Therefore, SD decoders do not allow for a one-to-one mapping between BER_{pre} and BER_{post}. Furthermore, modern digital coherent optical systems use joint designs of modulation and FEC, in a process known as coded modulation (CM) [5]. Among several CM schemes, popular choices are trellis-coded modulation (TCM) [5], multilevel coding (MLC) [6], non-binary coded modulation (NB-CM) [7], and bit-interleaved coded modulation (BICM) [8]. CM schemes are classified according to several architectural choices, and are usually evaluated by information-theoretical metrics. In this chapter we review conventional and information-theoretic metrics for several CM architectures.

8.2   Performance Evaluation with BER$_{pre}$ and the FEC Limit

Early optical communications systems based on on–off keying (OOK) and HD-FEC used BER_{pre} as a performance metric, given by

$$\text{BER}_{pre} = \frac{N_{be}}{N_{tb}}, \tag{8.1}$$

where N_{be} is the number of bit errors and N_{tb} is the number of transmitted bits. By the law of large numbers, BER_{pre} tends to the bit error probability as N_{tb} tends to infinity.

Until recently, the most popular performance evaluation approach in optical systems with FEC has been to estimate BER_{pre} and, using a one-to-one mapping related to the FEC performance, estimate BER_{post}. An implicit assumption in such method is that different systems with the same BER_{pre} obtain similar BER_{post} when using the same coding scheme [9]. This is a reasonable assumption, as errors in optical transmission systems are usually uncorrelated, or can be decorrelated after deinterleaving. Figure 8.1 illustrates the relationship between BER_{pre} and BER_{post}

---

[1]HD decoders can also exploit reliability information [2–4]. However, we consider here standard HD decoders designed to minimize metrics based on the Hamming distance.

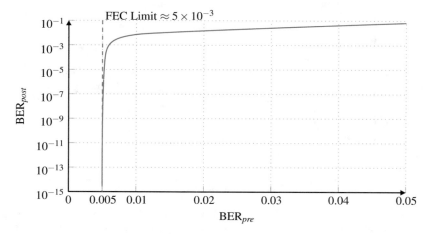

Fig. 8.1   Illustration of BER_{post} as a function of BER_{pre} for a hypothetical FEC scheme with FEC limit of approximately 5×10^{-3}. As BER_{pre} approaches the FEC limit, BER_{post} decreases abruptly

for a hypothetical coding scheme. Most FEC codes have a very abrupt curve, such that BER_{post} practically goes to zero after a certain threshold of low BER_{pre}. This threshold, called FEC Limit, is typically in the order of 10^{-2} or 10^{-3} (in Fig. 8.1, the FEC Limit is set to 0.005). The computational or experimental evaluation of optical communications systems based on BER_{pre} is basically limited to ensuring that it is below the FEC Limit.

Analyses based on a pre-FEC symbol error rate (SER_{pre}) are also possible, but less common. The SER_{pre} can be estimated in simulations and experiments as the average number of errors between the transmitted symbol sequence and the symbol sequence obtained after decision.

8.3   Performance Evaluation with Information-Theoretic Approaches

While the first generations of FEC codes deployed in optical communications systems used HD, modern coherent optical systems use more advanced SD schemes. In communications systems with SD-FEC, BER_{pre} is an insufficient metric to predict BER_{post} [3, 9, 10]. Furthermore, approaches using the FEC Limit do not support the evaluation of less conventional CM techniques, such as those using constellation shaping (CS) [11–14]. Alternatively, the optical communications community started to consider information-theoretic metrics, such as achievable information rates (AIRs). AIRs provide a limit on the amount of information bits per channel use that can be transmitted with arbitrarily low error probability through a channel [15], under constraints of modulation format and coding scheme.

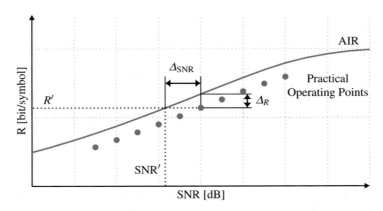

Fig. 8.2 Illustration of the AIR of a hypothetical CM scheme (solid orange curve) and the information rates achieved by practical implementations of the same CM scheme (blue dots). Practical implementations exhibit an SNR gap Δ_{SNR} and an information rate gap Δ_{R}

AIRs are achieved under ideal FEC schemes, with sufficiently long (and eventually impractical) codewords.

In practice, FEC codes exhibit a gap to ideal performance due to the use of pragmatic configurations, e.g., codewords of limited length and suboptimal decoding strategies intended to reduce implementation complexity [16]. Therefore, just as we used the BER$_{pre}$ FEC limit to predict BER$_{post}$ in HD-FEC, AIRs can provide a pre-FEC performance indicator to ensure an arbitrarily low BER$_{post}$ based on pre-calculated gaps related to specific FEC implementations. Figure 8.2 illustrates the AIR (in bit per symbol) for a hypothetical CM scheme (solid orange curve), and the information rate (also in bit per symbol) of practical implementations of the same CM scheme (blue dots). For the sake of simplicity, the figure assumes a constant SNR gap Δ_{SNR} and a constant information rate gap Δ_{R} for all operating points [17]. Suppose that an information rate R' with arbitrarily low BER$_{post}$ is desired. If an ideal FEC scheme was available, an SNR = SNR$'$ would be sufficient to ensure R'. If, however, practical FEC schemes are considered, the system must be designed to operate at SNR = SNR$' + \Delta_{\mathrm{SNR}}$, resulting in an AIR of $R' + \Delta_R$. The use of AIRs as pre-FEC performance metrics is discussed in more detail in the next sections.

8.3.1 Coded Modulation and AIRs

AIRs can be used as pre-FEC performance indicators for practical CM schemes. Therefore, in this section we review the classes of coded-modulation schemes and the AIRs applicable to each of them. Figure 8.3 shows the block diagram of a system using a generic CM transmission scheme. At the transmit side the CM

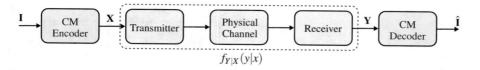

Fig. 8.3 Block diagram of a system using a generic CM transmission scheme. The CM encoder takes as input a binary information sequence **I** and generates a symbol sequence **X**. After the channel, the received sequence **Y** is processed by the CM decoder, which estimates the transmitted binary information sequence **Î**. Channel $f_{Y|X}(y|x)$ includes the transmitter, physical medium propagation channel (e.g., the optical channel), and the receiver. DSP stages are considered part of the transmitter and receiver

encoder receives a binary information sequence **I** and generates a symbol sequence **X**, including redundancy required for receive-side error correction.[2] Each element of **X** belongs to an alphabet of symbols $\mathcal{X}$ of size M. In practical systems, the sequence **X** is fed into the transmitter, which in optical systems include DSP algorithms, the laser, and the modulator. The signal produced by the transmitter is sent through the physical medium, in our case an optical link consisting of optical fibers, amplifiers, and passive devices. The digital coherent optical receiver includes the receiver front-end and the chain of DSP algorithms. The received symbol sequence **Y** is processed by the CM decoder, which estimates the transmitted binary information sequence, **Î**. Under the information-theoretical perspective, the channel described by the conditional probability density function (pdf) $f_{Y|X}(y|x)$ includes all stages between the generation of **X** and the detection of **Y**.

CM schemes can be implemented in several ways. At the CM encoder, the process of converting the binary information sequence **I** into the sequence **X** is commonly split into two steps. The binary information sequence is encoded using a FEC encoder supporting a binary or a non-binary FEC code. The FEC encoder transforms the information sequence into a coded sequence, called codeword. The set of all possible codewords is called a codebook. The codeword is mapped into symbols of a constellation using a memoryless mapper. At the CM decoder, the process of estimating the transmitted binary information sequence **Î** from the received symbol sequence **Y** is also commonly split into two stages. The demapper produces metrics related to **Y**. These metrics are sent to the FEC decoder, which estimates the transmitted codeword and, subsequently, the transmitted binary information sequence, generating estimate **Î**. CM schemes can be classified according to the FEC encoder/decoder implementation and its interaction with the mapper/demapper. CM decoders can use symbol-wise (SW) or bit-wise (BW) demappers. SW demappers are used in the case of non-binary FEC codes, while BW demappers are generally

---

[2] Uppercase letters (e.g., X) indicate random variables, with their realizations being indicated by lower case letters (e.g., x). The alphabets are indicated by calligraphic letters (e.g., $\mathcal{X}$). Boldface uppercase letters (e.g., **X**) indicate vectors of random variables. Boldface lowercase letters (e.g., **x**) indicate vectors of realizations of random variables.

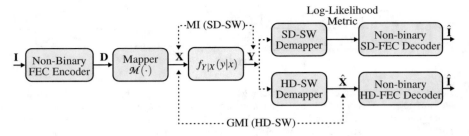

Fig. 8.4 CM scheme with non-binary FEC codes. The non-binary FEC scheme encodes a binary input sequence **I** into a non-binary codeword **D**. Codeword **D** is then mapped into the transmitted symbol sequence **X**. After the channel, the received sequence **Y** is forwarded to the CM decoder. In the case of SD decoding, a SD-SW demapper calculates log-likelihood metrics that are sent to a non-binary SD-FEC decoder. In the case of HD decoding, a HD-SW demapper generates estimates $\hat{\mathbf{X}}$ that are sent to a non-binary HD-FEC decoder. The MI is a relevant metric for SD-SW decoders, while the GMI is a relevant metric for HD-SW decoders (with Hamming distance metric)

used with binary FEC codes. In addition, decoders can be based on soft-decision (SD) or hard-decision (HD).

Figure 8.4 shows a block diagram of a generic CM scheme with non-binary FEC. A binary information sequence **I** is encoded by a non-binary FEC encoder into a non-binary codeword **D**. The codeword is then mapped into the transmitted sequence **X**, using a memoryless mapper $\mathcal{M}(\cdot)$ that operates on each element of **D**. Usually, the alphabet of the transmitted symbol constellation has the same size as the alphabet of the non-binary FEC code. For soft-decision symbol-wise (SD-SW) decoders, a SD-SW demapper calculates log-likelihood metrics (e.g., LLRs) using the received symbol sequence **Y**. The calculated values are sent to a non-binary soft-decision FEC (SD-FEC) decoder, which generates at its output estimate $\hat{\mathbf{I}}$. For hard-decision symbol-wise (HD-SW) decoders, a HD-SW demapper makes hard decisions on each symbol of the received sequence **Y**, producing an estimate of the transmitted sequence $\hat{\mathbf{X}}$. A non-binary hard-decision FEC (HD-FEC) decoder then generates estimate $\hat{\mathbf{I}}$. Examples of schemes with the aforementioned architectures are NB-CM schemes using (non-binary) low-density parity-check (LDPC) codes and SD decoders [3, 18], and NB-CM schemes with (non-binary) Staircase codes and HD decoders [4, 19].[3]

Figure 8.5 shows a block diagram of a generic CM scheme with binary FEC. A binary FEC encoder receives the information sequence **I** and encodes it into a binary codeword **D**. Codeword **D** is then mapped into the transmitted sequence **X** in

---

[3]Some HD decoders for Staircase codes use strategies based on the Hamming distance metric [4]. The Hamming distance between two sequences **A** and $\hat{\mathbf{A}}$ is the number of elements in which they differ from each other.

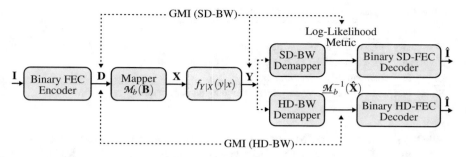

Fig. 8.5 CM scheme with binary FEC codes. The binary FEC scheme encodes a binary input sequence **I** into a binary codeword **D**. Codeword **D** is then mapped into the transmitted sequence **X**. After the channel, the received sequence **Y** is forwarded to the CM decoder. In the case of SD decoding, a SD-BW demapper calculates log-likelihood metrics (e.g., LLRs) that are sent to a binary SD-FEC decoder. In the case of HD decoding, a HD-BW demapper generates estimates of the binary label of symbols of the sequence $\hat{\mathbf{X}}$ and sends them to a binary HD-FEC decoder. The GMI is a relevant metric for both SD-BW decoders and HD-BW decoders (with Hamming distance metric)

sets of bits $\mathbf{B} = \{B_1, B_2, \cdots, B_m\}$, where $m = \log_2\{M\}$. Usually, a memoryless mapper $\mathcal{M}_b(\cdot)$ carries out a one-to-one mapping between $\mathbf{B}$ and elements of $\mathbf{X}$. For soft-decision bit-wise (SD-BW) decoders, log-likelihood metrics are calculated using the received sequence symbol sequence $\mathbf{Y}$. Unlike non-binary CM schemes, in this case LLRs are calculated with binary metrics. Finally, a binary SD-FEC decoder produces an estimate of the transmitted sequence $\hat{\mathbf{I}}$ using these soft values. Generally, the decoding process treats bits $\{B_1, B_2, \cdots, B_m\}$ as independent. For hard-decision bit-wise (HD-BW) decoders, the binary labels of the transmitted symbols are estimated using the received sequence $\mathbf{Y}$. Usually, hard decisions are made on the received symbols and their binary labels are obtained as $\mathcal{M}_b^{-1}(\hat{\mathbf{X}})$, where $\mathcal{M}_b^{-1}(\cdot)$ indicates the inverse mapping function. A binary HD-FEC then uses these binary labels to produce estimate $\hat{\mathbf{I}}$. Examples of schemes with the aforementioned structures are BICM schemes using LDPC codes and SD decoders [20], and CM schemes using binary Staircase codes and HD decoders [4].[4]

Figures 8.4 and 8.5 also show the relevant AIRs for the different CM schemes and between different points of the coding process. The MI is a relevant metric for CM schemes with non-binary FEC codes and SD-SW decoders. The GMI is a relevant metric for CM schemes with binary FEC codes and SD-BW decoders. The GMI is also a relevant metric for HD CM schemes with both non-binary and binary FEC codes, as long as decoding is based on the Hamming distance metric.

---

[4]BICM schemes have bit interleaving. In certain codes, e.g., for low-density parity-check (LDPC) codes, interleaving can be implicit in the parity-check matrix.

Table 8.1 CM schemes and their relevant AIRs (also shown in Figs. 8.4 and 8.5)

CM scheme	Relevant AIR
Non-binary FEC code with SD-SW decoding	MI
Non-binary FEC code with HD-SW decoding (Hamming distance metric)	GMI
Binary FEC code with SD-BW decoding	GMI
Binary FEC code with HD-BW decoding (Hamming distance metric)	GMI

Table 8.1 summarizes the CM schemes and their relevant performance metrics.[5] The next sections discuss in detail the calculation of the MI and the GMI for different CM schemes.

8.3.2   Mutual Information

Let us first consider that X and Y are complex discrete-valued random variables that represent the input and the output of a channel. X takes values from an alphabet of symbols $X = \{x_1, x_2, \cdots, x_M\}$ of size $|X| = M$ according to the probability mass function (pmf) $p_X(x) = Pr\{X = x\}$, $x \in X$. Similarly, Y takes values from an alphabet of symbols $Y = \{y_1, y_2, \cdots, y_M\}$ of size $|Y| = M$ according to the pmf $p_Y(y) = Pr\{Y = y\}$, $y \in Y$. We call $p_{Y|X}(y|x)$ the conditional pmf of Y given X, indicating the transition probabilities of a discrete-time memoryless channel. The average information or similarly the average uncertainty associated with channel input X is measured by its entropy, which is defined as [23]

$$H_e(X) = \mathbb{E}\{-\log_2[p_X(x)]\}, \qquad (8.2)$$

where $\mathbb{E}\{\cdot\}$ indicates expectation. As the logarithm in (8.2) is taken in base 2, $H_e(X)$ is measured in bits. Taking into account that X is a discrete random variable, (8.2) becomes

$$H_e(X) = -\sum_{i=1}^{M} p_X(x_i) \log_2 [p_X(x_i)]. \qquad (8.3)$$

Extending the idea of entropy, it is also possible to measure the amount of information about channel input X, given a certain realization y_1 of channel Y [24],

$$H_e(X|Y = y_1) = -\sum_{i=1}^{M} p_{X|Y}(x_i|y_1) \log_2 \left[p_{X|Y}(x_i|y_1)\right], \qquad (8.4)$$

---

[5]Some CM schemes with binary FEC codes also have the MI as a relevant metric. These schemes are out of the scope of this book. Some examples are BICM schemes that employ iterations between the SD-BW demapper and the binary SD-FEC decoder, and MLC schemes with multistage decoding. For more details, we refer the reader to [21] and [22].

where $p_{X|Y}(x_i|y_1)$ is the conditional pmf of the channel input $X = x_i$ given that the channel output is $Y = y_1$. Assuming all possible outcomes of Y, the average information in the channel input X given that the channel output Y was observed can be measured by the conditional entropy $H_e(X|Y)$ as

$$H_e(X|Y) = -\sum_{i=1}^{M}\sum_{j=1}^{M} p_{X,Y}(x_j, y_i) \log_2 \left[p_{X|Y}(x_j|y_i) \right], \tag{8.5}$$

where $p_{X,Y}(x_j, y_i)$ is the joint pmf of the channel input $X = x_j$ and the channel output $Y = y_i$.

The MI between X and Y is defined as [23]

$$I(X; Y) = H_e(X) - H_e(X|Y). \tag{8.6}$$

The MI indicates the reduction in average information associated with the channel input X given that the channel output Y was observed, measured in bit per symbol (or bit per channel use). Applying the expressions for $H_e(X)$ and $H_e(X|Y)$ into (8.6) yields

$$I(X; Y) = -\sum_{l=1}^{M} p_X(x_l) \log_2 [p_X(x_l)] + \sum_{i=1}^{M}\sum_{j=1}^{M} p_{X,Y}(x_j, y_i) \log_2 \left[p_{X|Y}(x_j|y_i) \right]. \tag{8.7}$$

Using the law of total probability, and the Bayes' theorem, yields

$$I(X; Y) = \sum_{j=1}^{M}\sum_{i=1}^{M} p_{X,Y}(x_j, y_i) \log_2 \left[\frac{p_{Y|X}(y_i|x_j)}{p_Y(y_i)} \right] \tag{8.8}$$

$$= \sum_{j=1}^{M} p_X(x_j) \sum_{i=1}^{M} p_{Y|X}(y_i|x_j) \log_2 \left[\frac{p_{Y|X}(y_i|x_j)}{\sum_{l=1}^{M} p_X(x_l) p_{Y|X}(y_i|x_l)} \right]. \tag{8.9}$$

A more compact expression for the MI in (8.8) and (8.9) is

$$I(X; Y) = \mathbb{E}_{X,Y} \left\{ \log_2 \left[\frac{p_{Y|X}(y|x)}{p_Y(y)} \right] \right\}, \tag{8.10}$$

where $\mathbb{E}_{X,Y}\{\cdot\}$ denotes expectation over the joint distribution of X and Y.

Let us now consider that the channel output Y is a complex continuous random variable with probability density function (pdf) $f_Y(y)$. In this case, the pmf $p_{Y|X}(y|x)$ in (8.9) is replaced by a pdf $f_{Y|X}(y|x)$, and the inner summation is replaced by an integral over the support set of Y, $\mathcal{Y}$. The MI is then given by

$$I(X; Y) = \sum_{j=1}^{M} p_X(x_j) \int_{\mathcal{Y}} f_{Y|X}(y|x_j) \log_2 \left[\frac{f_{Y|X}(y|x_j)}{\sum_{l=1}^{M} p_X(x_l) f_{Y|X}(y|x_l)} \right] dy.$$

(8.11)

For the scenario where both X and Y are complex continuous random variables, all pmfs in (8.8) are replaced by pdfs, and summations are replaced by integrals over the support set of complex random variables X and $\mathcal{Y}$. Thus, the MI becomes

$$I(X; Y) = \int_X \int_{\mathcal{Y}} f_{X,Y}(x, y) \log_2 \left[\frac{f_{Y|X}(y|x)}{f_Y(y)} \right] dxdy.$$

(8.12)

Similarly to (8.10), the MI in (8.11) and (8.12) can be expressed for the continuous case as

$$I(X; Y) = \mathbb{E}_{X,Y} \left\{ \log_2 \left[\frac{f_{Y|X}(y|x)}{f_Y(y)} \right] \right\}.$$

(8.13)

Shannon's channel coding theorem gives an operational meaning of AIR to the MI [15]. It states that, if transmission occurs at rate $R < I(X, Y)$, in bit per symbol (or bit per channel use), there is a channel code that ensures an arbitrarily small error probability after optimum decoding.[6]

Maximizing $I(X; Y)$ for a given channel over all possible input distributions gives the channel capacity C. Supposing the scenario with discrete random variables at the input and the output, the capacity of a memoryless channel is given by

$$C = \max_{p_X(x)} \{I(X; Y)\}.$$

(8.14)

Assuming continuous random variables at input and output, the capacity of a memoryless channel with input power constraint $\mathbb{E}\{|X|^2\} \le P$ is

$$C = \max_{f_X(x):\mathbb{E}\{|X|^2\} \le P} \{I(X; Y)\}.$$

(8.15)

For a complex additive white Gaussian noise (AWGN) channel, $y = x + \eta$, for which η has a zero-mean complex Gaussian distribution with variance $2\sigma_\eta^2$, the conditional pdf becomes

$$f_{Y|X}(y|x) = \frac{1}{2\pi\sigma_\eta^2} e^{-|y-x|^2/(2\sigma_\eta^2)}.$$

(8.16)

---

[6]The rate R can be interpreted as the rate of a coded modulation scheme.

In this case, the MI is maximized if input X is also complex and Gaussian distributed. If X has zero mean and variance P, the channel capacity is given by Shannon [15]

$$C = \log_2 \left(1 + \frac{P}{2\sigma^2} \right). \tag{8.17}$$

This is the well-known Shannon capacity formula for the AWGN channel.

The MI is an AIR for non-binary CM schemes with optimal SD-SW decoders,[7] i.e.,

$$I_{\text{SD-SW}} = I(X; Y). \tag{8.18}$$

Assuming equally likely symbols ($p_X(x_i) = 1/M$), (8.18) reduces to

$$I_{\text{SD-SW}} = m + \frac{1}{M} \sum_{i=1}^{M} \int_{\mathcal{Y}} f_{Y|X}(y|x_i) \log_2 \left[\frac{f_{Y|X}(y|x_i)}{\sum_{j=1}^{M} f_{Y|X}(y|x_j)} \right] dy. \tag{8.19}$$

In simulations and experiments, estimates of $I_{\text{SD-SW}}$ are obtained as [25, 26]

$$I_{\text{SD-SW}} \approx \frac{1}{N_s} \sum_{k=1}^{N_s} \log_2 \left[\frac{f_{Y|X}(y[k]|x[k])}{\frac{1}{M} \sum_{j=1}^{M} f_{Y|X}(y[k]|x_j)} \right], \tag{8.20}$$

where $x[k]$ are the transmitted symbols and $y[k]$ are the symbols at the output of the DSP chain, for $k = \{1, \cdots, N_s\}$, where N_s is the number of transmitted/received symbols. The accuracy of this estimate is directly related to N_s [26].

Assuming the circularly symmetric AWGN channel, $I_{\text{SD-SW}}$ becomes [27]

$$I_{\text{SD-SW}}^{\text{AWGN}} = m - \frac{1}{M} \sum_{i=1}^{M} \int_{\mathcal{H}} f_H(\eta) \log_2 \sum_{j=1}^{M} e^{-(|x_i-x_j|^2 + 2\mathbb{R}\{\eta(x_i-x_j)\})/(2\sigma_\eta^2)} d\eta, \tag{8.21}$$

where η is the realization of a Gaussian random variable H with zero mean and variance $2\sigma_\eta^2$ and $\mathbb{R}\{\cdot\}$ takes the real component of $\{\cdot\}$.

In computational calculations, the integral in (8.21) can be evaluated by the Gauss–Hermite (GH) quadrature approximation [28], yielding [27]

---

[7] As previously discussed, the MI is also a relevant metric for some binary CM schemes, e.g., BICM schemes with iterative decoding and MLC schemes with multi-stage decoding.

Table 8.2 Nodes ζ and weights ψ of a $J = 10$ point GH quadrature approximation [28]

p_1 ; p_2	ζ (Nodes)	ψ (Weights)
1	-3.43615911883773760	$0.76404328552326206 \times 10^{-5}$
2	-2.53273167423278980	$0.13436457467812327 \times 10^{-2}$
3	-1.75668364929988177	$0.33874394455481063 \times 10^{-1}$
4	-1.03661082978951365	0.24013861108231469
5	-0.34290132722370461	0.61086263373532580
6	0.34290132722370461	0.61086263373532580
7	1.03661082978951365	0.24013861108231469
8	1.75668364929988177	$0.33874394455481063 \times 10^{-1}$
9	2.53273167423278980	$0.13436457467812327 \times 10^{-2}$
10	3.43615911883773760	$0.76404328552326206 \times 10^{-5}$

$$
I_{\text{SD-SW}}^{\text{AWGN}} \approx m - \frac{1}{\pi M} \sum_{i=1}^{M} \sum_{p_1=1}^{J} \sum_{p_2=1}^{J} \psi_{p_1} \psi_{p_2} \cdot
$$

$$
\times \log_2 \sum_{j=1}^{M} e^{-\left(|x_i - x_j|^2 + 2\sqrt{2}\sigma_\eta \mathbb{R}\{(\zeta_{p_1} + j\zeta_{p_2})(x_i - x_j)\} \right) / \left(2\sigma_\eta^2 \right)}, \tag{8.22}
$$

where ψ and ζ are the weights and nodes of a J-point GH quadrature approximation. Table 8.2 shows ψ and ζ for $J = 10$ [28]. Figure 8.6 shows $I_{\text{SD-SW}}^{\text{AWGN}}$ for 16-QAM and QPSK modulation formats with uniform symbol probabilities (solid lines), obtained by (8.22) with $J = 10$. The figure also shows the maximum AIR considering all possible input distributions, which is given by the Shannon capacity formula in (8.17). It is interesting to note that the AIRs obtained for 16-QAM and QPSK in Fig. 8.6 do not cross. Therefore, in this idealized scenario, higher-order modulation formats are always advantageous over lower-order modulation formats, in the sense that they provide a higher AIR for a given SNR [5].

The AIRs of (8.18) and (8.19) assume that the FEC decoder knows the exact channel law $f_{Y|X}(y|x)$. However, $f_{Y|X}(y|x)$ may be unknown in certain scenarios. In these cases, it is common to assume an auxiliary channel at the receiver with pdf $g_{Y|X}(y|x)$. This decoding process is called *mismatched*, referring to the fact that decoding is performed assuming a channel law $g_{Y|X}(y|x)$ that is an approximation to the actual channel law $f_{Y|X}(y|x)$. Under mismatched decoding, a lower bound on the AIR is given by Arnold et al. [25] and Fehenberger et al. [29]

$$
\tilde{I}_{\text{SD-SW}} = m + \frac{1}{M} \sum_{i=1}^{M} \int_{\mathcal{Y}} f_{Y|X}(y|x_i) \log_2 \left[\frac{g_{Y|X}(y|x_i)}{\sum_{j=1}^{M} g_{Y|X}(y|x_j)} \right] dy, \tag{8.23}
$$

where the tilde in $\tilde{I}_{\text{SD-SW}}$ indicates the use of an auxiliary channel. The mismatched decoding process leads to a rate loss due to the use of a suboptimal decoding metric $g_{Y|X}(y|x)$ [25], i.e.,

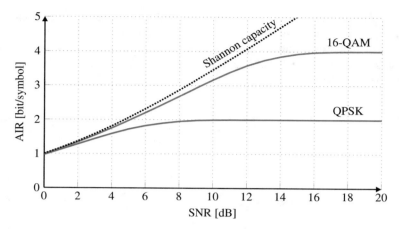

Fig. 8.6 AIRs $I_{\text{SD-SW}}^{\text{AWGN}}$ (solid lines) as a function of SNR for non-binary CM schemes with SD-SW decoders and ideal FEC (8.21). An AWGN channel is assumed. The Shannon capacity is indicated by the dotted line

$$\tilde{I}_{\text{SD-SW}} \leq I_{\text{SD-SW}}, \tag{8.24}$$

where equality holds if $f_{Y|X}(y|x) = g_{Y|X}(y|x)$.

Until now we considered that the channel is memoryless, which does not hold for certain regimes of nonlinear propagation. Approaching the AIR for a channel with memory also requires a decoder that takes into account the channel memory. However, as the exact channel law may be hard to find, AIR calculation usually resorts to a memoryless auxiliary channel with pdf $g_{Y|X}(y|x)$ having a circularly symmetric Gaussian distribution [2, 29]. This approach also results in a lower bound on the AIR, i.e.,

$$\tilde{I}_{\text{SD-SW}} \leq I_{memory}. \tag{8.25}$$

In simulations and experiments, $\tilde{I}_{\text{SD-SW}}$ can be computationally estimated as

$$\tilde{I}_{\text{SD-SW}} \approx \frac{1}{N_s} \sum_{k=1}^{N_s} \log_2 \left[\frac{g_{Y|X}(y[k]|x[k])}{\frac{1}{M} \sum_{j=1}^{M} g_{Y|X}(y[k]|x_j)} \right]. \tag{8.26}$$

It is common to model the auxiliary channel as an AWGN channel. In this case, $g_{Y|X}(y|x)$ is given by Buchali [30]

$$g_{Y|X}(y|x) = \frac{1}{2\pi\sigma_\eta^2} e^{-|y-hx|^2/(2\sigma_\eta^2)}, \tag{8.27}$$

where σ_η^2 is the noise variance per dimension and h is a multiplicative scalar value. Considering h is important because experiments and simulations give outputs that are approximately the sum of signal and noise, without discriminating the signal and noise powers. Estimates of h and σ_η^2 can be generated from the transmitted and received constellations as

$$\hat{h} = \frac{\sum_{k=1}^{N_s} x^*[k]y[k]}{\sum_{k=1}^{N_s} |x[k]|^2}, \tag{8.28}$$

and

$$\hat{\sigma}_\eta^2 = \frac{1}{2N_s} \sum_{k=1}^{N_s} \left| y[k] - \hat{h}x[k] \right|^2. \tag{8.29}$$

8.3.3   Generalized Mutual Information

The MI is the AIR considering constraints on the physical channel and on the modulation format, assuming ideal FEC (or an ideal non-binary FEC scheme with SD and SW decoding). Other CM architectures, e.g., with non-binary FEC with HD decoding and binary FEC with HD and SD decoding, require alternative methods to assess their AIRs. In fact, these schemes are commonly treated under the mismatched decoding perspective, where the GMI is a useful metric. Assuming a generic decoding metric $q(x, y)$, the GMI (in bit per symbol) can be expressed as [21, 31]

$$\tilde{I}_{\mathrm{GMI}}(X; Y) = \max_{a \geq 0} \mathbb{E}_{X,Y} \left\{ \log_2 \left[\frac{[q(x, y)]^a}{\sum_{j=1}^{M} p_X(x_j)[q(x_j, y)]^a} \right] \right\}. \tag{8.30}$$

The tilde on $\tilde{I}_{\mathrm{GMI}}$ indicates the use of a generic decoding metric $q(x, y)$, analogous to the case of AIRs for SD-SW decoders with an auxiliary channel.[8]

AIRs for Non-Binary CM Schemes with HD-SW Decoders

The GMI $\tilde{I}_{\mathrm{GMI}}(X; \hat{X})$, between X and its estimate $\hat{X}$, generated from the received symbol Y after hard decision, is an AIR for non-binary CM schemes with HD-

---

[8]Note that, if $q(x, y) = f_{Y|X}(y|x)$ and $a = 1$, (8.30) reduces to (8.13), where $f_Y(y) = \sum_{j=1}^{M} p_X(x_j) f_{Y|X}(y, x_j)$.

SW decoders based on the Hamming distance metric [4, 32].[9] For equally likely symbols, $I_{\text{HD-SW}}$ is given by

$$I_{\text{HD-SW}} = \tilde{I}_{GMI}(X; \hat{X}) = m - H_2(\varepsilon_{\text{SW}}) - \varepsilon_{\text{SW}} \log_2(M - 1), \tag{8.31}$$

where $\varepsilon_{\text{SW}} = Pr\{\hat{X} \neq x | X = x\}$ is the uncoded symbol error probability, and $H_2(\varepsilon_{\text{SW}})$ is the binary entropy function of ε_{SW} [4]

$$H_2(\varepsilon_{\text{SW}}) = -(1 - \varepsilon_{\text{SW}}) \log_2(1 - \varepsilon_{\text{SW}}) - \varepsilon_{\text{SW}} \log_2(\varepsilon_{\text{SW}}). \tag{8.32}$$

It is interesting to observe that the AIR calculated by (8.31) is equivalent to that of an M-ary symmetric channel with error probability ε_{SW} [33, 34]. Theoretical expressions for $I_{\text{HD-SW}}$ can be calculated resorting to classic symbol error probability formulas. For example, assuming a square M-QAM format and an AWGN channel, ε_{SW} is given by Goldsmith [35]

$$\varepsilon_{\text{SW}} = 1 - \left[1 - \frac{\sqrt{M} - 1}{\sqrt{M}} \text{erfc} \left(\sqrt{\frac{3\text{SNR}}{2(M - 1)}} \right) \right]^2. \tag{8.33}$$

Likewise, in simulations and experiments, estimates of $I_{\text{HD-SW}}$ can be generated using the SER obtained experimentally.

Figure 8.7 shows $I_{\text{HD-SW}}$ for the AWGN channel, obtained with (8.31), using SER values given by (8.33). The solid lines indicate the performance of the 16-QAM and QPSK modulation formats, whereas the dotted line indicates the Shannon capacity. As hard-decision decoding neglects soft channel information, the AIRs obtained with HD-SW decoders are lower than those for SD-SW decoders shown in Fig. 8.6. Unlike the case for SD-SW decoders, the QPSK performance surpasses that of 16-QAM at low SNR values.

AIRs for Binary CM Schemes with SD-BW Decoders

Binary CM schemes with SD-BW decoders usually use a decoding metric $q(\mathbf{b}, y)$ in the form of a product of metrics calculated with the individual bits $\{B_1, B_2, \cdots, B_m\}$. The exact AIR for this configuration is an open research topic, but the GMI proved to be a valid reference [9]. In this case, the GMI in (8.30) can be expressed as [9]

---

[9]Under the theory of mismatched decoding, we can use the GMI expression to obtain an AIR for non-binary CM schemes with HD-SW decoders based on the Hamming distance metric.

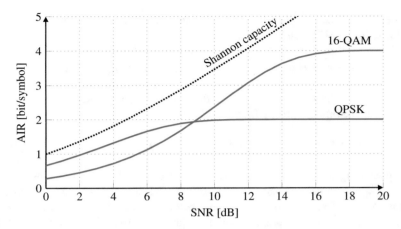

Fig. 8.7 AIRs $I_{\text{HD-SW}}$ (solid lines) as a function of SNR for non-binary CM schemes with HD-SW decoders (based on the Hamming distance metric) and ideal FEC (8.31). An AWGN channel is assumed. The Shannon capacity is indicated by the dotted line

$$\tilde{I}_{\text{GMI}}(\mathbf{B}; Y) = \max_{a \geq 0} \mathbb{E}_{\mathbf{B}, Y} \left\{ \log_2 \left[\frac{[q(\mathbf{b}, y)]^a}{\sum_{\mathbf{b}' \in \{0,1\}^m} p_{\mathbf{B}}(\mathbf{b}')[q(\mathbf{b}', y)]^a} \right] \right\}, \tag{8.34}$$

where $\mathbf{B} = \{B_1, B_2, \cdots, B_m\}$, $m = \log_2 M$, is the set of bits mapped into X using a memoryless one-to-one mapper $\mathcal{M}(\cdot)$.

For uniformly distributed bits, a decoding metric can be defined in terms of the product of conditional probabilities of the channel output Y given bit B_l, $f_{Y|B_l}(y|b_l)$, $l = \{1, 2, \cdots, m\}$

$$f_{Y|B_l}(y|b_l) = \sum_{i \in \mathcal{I}_b^l} f_{Y|X}(y|x_i), \tag{8.35}$$

where $\mathcal{I}_b^l$ represents the set of indices of constellation symbols $x \in X$ whose bit at position l is $b_l = b \in \{0, 1\}$. The GMI in (8.34) becomes [27]

$$I_{\text{SD-BW}} = m + \frac{1}{M} \sum_{l=1}^{m} \sum_{b \in \{0,1\}} \sum_{i \in \mathcal{I}_b^l} \int_{\mathcal{Y}} f_{Y|X}(y|x_i) \log_2 \left[\frac{\sum_{z \in \mathcal{I}_b^l} f_{Y|X}(y|x_z)}{\sum_{j=1}^{M} f_{Y|X}(y|x_j)} \right] dy. \tag{8.36}$$

Assuming an AWGN channel, $I_{\text{SD-BW}}$ is given by Alvarado et al. [27]

$$I_{\text{SD-BW}}^{\text{AWGN}} = m - \frac{1}{M} \sum_{l=1}^{m} \sum_{b \in \{0,1\}} \sum_{i \in \mathcal{I}_b^l} \int_{\mathcal{H}} f_H(\eta) \cdot$$

$$\times \log_2 \left[\frac{\sum_{j=1}^{M} e^{-(|x_i - x_j|^2 + 2\mathbb{R}\{\eta(x_i - x_j)\})/(2\sigma_\eta^2)}}{\sum_{z \in \mathcal{I}_b^l} e^{-(|x_i - x_z|^2 + 2\mathbb{R}\{\eta(x_i - x_z)\})/(2\sigma_\eta^2)}} \right] d\eta. \tag{8.37}$$

In order to simplify computations, the FEC decoder usually uses log-likelihood ratios (LLRs) as decoding metrics, defined as

$$\gamma_l = \log \left[\frac{f_{Y|B_l}(y|b_l = 1)}{f_{Y|B_l}(y|b_l = 0)} \right]. \tag{8.38}$$

The expression for $I_{\text{SD-BW}}$ in terms of LLRs is largely simplified [36]

$$I_{\text{SD-BW}} = m - \frac{1}{2} \sum_{l=1}^{m} \sum_{b \in \{0,1\}} \int_{\Gamma} f_{\Gamma_l|B_l}(\gamma|b) \log_2 \left[1 + e^{(-1)^b \gamma} \right] d\gamma, \tag{8.39}$$

where $f_{\Gamma_l|B_l}(\gamma|b)$ is the conditional pdf of channel output in the form of LLRs Γ_l, given channel input B_l.

As discussed in the case of SD-SW decoders, $f_{Y|X}(y|x)$ and consequently $f_{Y|B_l}(y|b_l)$ may be unknown. In this case, the decoder uses an auxiliary channel. For a (generic) bit metric $q_l(b_l, y)$, (mismatched) LLRs are calculated as

$$\tilde{\gamma}_l = \log \left[\frac{q_l(b_l = 1|y)}{q_l(b_l = 0|y)} \right], \tag{8.40}$$

and an AIR can be obtained as

$$\tilde{I}_{\text{SD-BW}} = m - \min_{a \geq 0} \frac{1}{2} \sum_{l=1}^{m} \sum_{b \in \{0,1\}} \int_{\tilde{\Gamma}} f_{\tilde{\Gamma}_l|B_l}(\tilde{\gamma}|b) \log_2 \left[1 + e^{a(-1)^b \tilde{\gamma}} \right] d\tilde{\gamma}. \tag{8.41}$$

The GMIs of (8.39) and (8.41) are AIRs for binary CM schemes with SD-BW decoders that use decoding metrics obtained as the product of bit metrics $f_{Y|B_l}(y|b_l)$ and $q_l(b_l, y)$. As examples, we can mention BICM schemes without iterative decoding, and MLC schemes with parallel, independent decoding of the individual levels (PDL) [9, 21, 22]. The AIRs obtained with binary CM schemes and SD-BW decoders depend on the bit mapping scheme, and Gray mapping reduces the loss with respect to $I(X; Y)$ [36].

As in the case of SD-SW decoders, we can estimate the AIR for SD-BW decoders under the AWGN channel using the GH quadrature approximation. The rate $I_{\text{SD-BW}}^{\text{AWGN}}$ in (8.37) can be estimated as [27] (see Table 8.2 for ψ and ζ for $J = 10$ [28])

$$I_{\text{SD-BW}}^{\text{AWGN}} \approx m - \frac{1}{\pi M} \sum_{l=1}^{m} \sum_{b \in \{0,1\}} \sum_{i \in \mathcal{I}_b^l} \sum_{p_1=1}^{J} \sum_{p_2=1}^{J} \psi_{p_1} \psi_{p_2} \cdot$$

$$\times \log_2 \left[\frac{\sum_{j=1}^{M} e^{-\left(|x_i - x_j|^2 + 2\sqrt{2}\sigma_\eta \mathbb{R}\{(\zeta_{p_1} + j\xi_{p_2})(x_i - x_j)\} \right)/(2\sigma_\eta^2)}}{\sum_{z \in \mathcal{I}_b^l} e^{-\left(|x_i - x_z|^2 + 2\sqrt{2}\sigma_\eta \mathbb{R}\{(\zeta_{p_1} + j\xi_{p_2})(x_i - x_z)\} \right)/(2\sigma_\eta^2)}} \right].$$

$$(8.42)$$

In simulation and experiments, $I_{\text{SD-BW}}$ can be estimated, according to (8.39), as [27]

$$I_{\text{SD-BW}} \approx m - \frac{1}{N_s} \sum_{l=1}^{m} \sum_{k=1}^{N_s} \log_2 \left[1 + e^{(-1)^{b_l[k]} \gamma_l[k]} \right]. \qquad (8.43)$$

The lth LLR $\gamma_l[k]$ is calculated applying (8.38) to the kth symbol $y[k]$ at the output of the DSP chain. Term $b_l[k]$ is the lth bit of the binary label $\mathbf{b}[k] = \{b_1[k], b_2[k], \cdots, b_m[k]\}$, mapped into the transmitted symbol $x[k]$ at instant k. Analogously, for the LLRs in (8.40), the mismatched AIR $\tilde{I}_{\text{SD-BW}}$ in (8.41) becomes

$$\tilde{I}_{\text{SD-BW}} \approx m - \min_{a \geq 0} \frac{1}{N_s} \sum_{l=1}^{m} \sum_{k=1}^{N_s} \log_2 \left[1 + e^{a(-1)^{b_l[k]} \tilde{\gamma}_l[k]} \right]. \qquad (8.44)$$

Let us now consider an auxiliary channel having a circularly symmetric Gaussian distribution, as in (8.27). We can define $q_l(b_l, y) = g_{Y|B_l}(y|b_l)$, where $g_{Y|B_l}(y|b_l)$ is the auxiliary channel with output Y given input B_l. Using (8.40), the lth LLR for the kth received symbol is given by

$$\tilde{\gamma}_l[k] = \log \left[\frac{\sum_{i \in \mathcal{I}_1^l} e^{-|y[k] - hx_i|^2/(2\sigma_\eta^2)}}{\sum_{j \in \mathcal{I}_0^l} e^{-|y[k] - hx_j|^2/(2\sigma_\eta^2)}} \right], \qquad (8.45)$$

where h and σ_η^2 can be estimated from the data, as in (8.28) and (8.29). The LLRs calculated with (8.45) can then be applied in (8.44) to estimate $I_{\text{SD-BW}}^{\text{AWGN}}$.

Figure 8.8 shows $I_{\text{SD-BW}}^{\text{AWGN}}$ for the 16-QAM and QPSK modulation formats, assuming uniform symbol probabilities and Gray mapping (solid lines). The AIRs were obtained computing (8.42) with $J = 10$. The dotted line indicates the Shannon capacity (dotted line). Comparing the AIRs obtained with SD-SW decoding in Fig. 8.6 with those for SD-BW decoding in Fig. 8.8, we can observe a slight reduction for the 16-QAM format caused by the BW processing. For QPSK modulation both schemes exhibit the same performance. This is expected, as a QPSK constellation with Gray mapping can be decomposed into two BPSK constellations in quadrature [27, 36].

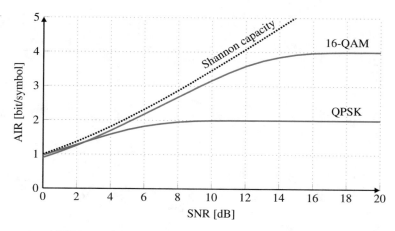

Fig. 8.8 AIRs $I_{\text{SD-BW}}^{\text{AWGN}}$ (solid lines) as a function of SNR for binary CM schemes with SD-BW decoders and ideal FEC (8.37). An AWGN channel is assumed. The Shannon limit is indicated by the dotted line

AIRs for Binary CM Schemes with HD-BW Decoders

Binary CM schemes with HD-BW decoding based on the Hamming distance metric also use the GMI as AIR. For equally likely symbols, the GMI in (8.30) becomes [4, 32]

$$I_{\text{HD-BW}} = m\left[1 - H_2(\varepsilon_{\text{BW}})\right],\tag{8.46}$$

where ε_{BW} is the uncoded bit error probability across the m bit positions [29, 37]

$$\varepsilon_{\text{BW}} = \frac{1}{m}\sum_{l=1}^{m} Pr\left(\hat{B}_l \neq b_l | B_l = b_l\right).\tag{8.47}$$

The GMI of (8.46) is an AIR for binary CM schemes combined with HD-BW decoding based on Hamming distance metric, under equally likely symbols. It is interesting to note that the AIR in (8.46) is equivalent to that of transmitting over m parallel binary symmetric channels with error probability ε_{BW} [34].

Assuming an AWGN channel and Gray mapping, theoretical $I_{\text{HD-BW}}$ values can be calculated by means of theoretical bit error probability expressions. For QPSK we have

$$\varepsilon_{\text{BW}} = \frac{1}{2}\text{erfc}\left(\sqrt{\frac{\text{SNR}}{2}}\right),\tag{8.48}$$

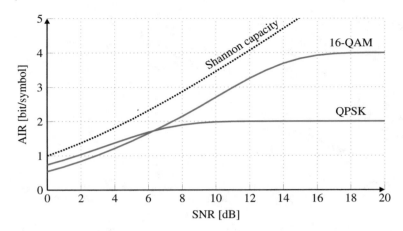

Fig. 8.9 AIRs $I_{\text{HD-BW}}$ (solid lines) as a function of SNR for binary CM schemes with HD-BW decoders (based on the Hamming distance metric) and ideal FEC (8.31). An AWGN channel is assumed. The Shannon capacity is indicated by the dotted line

and for 16-QAM [38]

$$\varepsilon_{\text{BW}} = \frac{3}{8}\text{erfc}\left(\sqrt{\frac{\text{SNR}}{10}}\right) + \frac{1}{4}\text{erfc}\left(3\sqrt{\frac{\text{SNR}}{10}}\right) - \frac{1}{8}\text{erfc}\left(5\sqrt{\frac{\text{SNR}}{10}}\right). \tag{8.49}$$

In simulations and experiments, $I_{\text{BW-HD}}$ can be estimated using the measured BER.

Figure 8.9 shows $I_{\text{HD-BW}}$ assuming an AWGN channel. The performance of the 16-QAM and QPSK modulation formats with uniform symbol probabilities and Gray mapping are evaluated (solid lines). The AIRs $I_{\text{HD-BW}}$ are obtained with (8.46), using as estimates of ε_{BW} the expressions in (8.48) and (8.49) for QPSK and 16-QAM. The Shannon capacity limit is indicated by the dotted line. The AIRs obtained with HD-BW decoders exhibit penalties when compared to that obtained with SD-BW decoders. On the other hand HD-BW decoders achieve higher rates than HD-SW decoders. This observation can be explained by the fact that decoding based on Hamming distance metrics does not provide reliability information related to individual transition probabilities [34].

8.3.4 Normalized Generalized Mutual Information

In certain cases it is convenient to use the normalized GMI (NGMI) as an AIR for binary CM schemes with SD-BW decoding. Assuming a uniform symbol distribution, the NGMI can be defined as

(a)   (b)

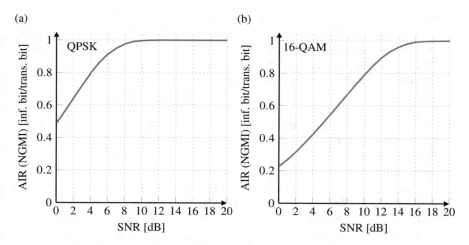

Fig. 8.10 NGMI $I_{\text{SD-BW}}^{\text{AWGN}}/m$ (solid lines) as a function of SNR for binary CM schemes with SD-BW decoders and ideal FEC, with (**a**) QPSK and (**b**) 16-QAM formats

$$\tilde{I}_{\text{NGMI}}(\mathbf{B}; Y) = \frac{\tilde{I}_{\text{GMI}}(\mathbf{B}; Y)}{m}. \qquad (8.50)$$

While the MI and the GMI indicate the number of information bits per transmitted symbol, the NGMI indicates the number of information bits conveyed by the channel per transmitted bit, and is limited to the range of [0,1].

Figure 8.10 shows AIRs in the form of NGMI for binary CM schemes with SD-BW decoders. The QPSK performance is shown in Fig. 8.10a, while the 16-QAM performance is shown in Fig. 8.10b. In both cases, uniform symbol probabilities and Gray mapping are considered. The AIRs are obtained computing (8.42) with $J = 10$, and dividing the result by $m = 2$ for QPSK, and by $m = 4$ for 16-QAM. Unlike previous figures, the results for the 16-QAM and QPSK modulation formats are plotted separately. This is because comparing the performance of different modulation formats through NGMIs are meaningful if the evaluated schemes result in the same gross (pre-FEC) bit rate. In this case, the NGMI evaluates the highest net information rate for a given gross transmitted bit rate.

8.4   Problems

1. Simulate the transmission of single-polarization Nyquist-shaped QPSK and 16-QAM signals at 50 GBaud. Perform pulse shaping at 16 Sa/Symbol using a root-raised cosine (RRC) filter with roll-off factor $\beta^{\text{RC}} = 0.1$ and a span of 20

symbols.[10] Consider OSNRs in the interval from 5 dB to 25 dB. At the receiver, assume matched filtering. Estimate the AIR for the scenarios with SD-SW and SD-BW decoders, assuming in both cases a circularly symmetric Gaussian channel. Present the results for the different modulation formats separately. Compare simulation and theoretical curves.

2. Repeat Problem 1, considering HD-SW and HD-BW decoding, for a decoder based on the Hamming distance.

3. Consider the scenario of Problem 1. Estimate the OSNR gap to Shannon capacity for the QPSK and 16-QAM formats. Assume SD-SW and SD-BW decoders. Plot the OSNR gap to Shannon capacity as a function of the OSNR. Compare simulation and theoretical results.

4. Repeat Problem 1, considering the transmission of polarization-multiplexed signals, with an OSNR in the interval from 8 dB to 25 dB.

5. Simulate the transmission of Nyquist-shaped QPSK signals with polarization multiplexing at 50 GBaud. Perform pulse shaping using the same parameters of Problem 1. Consider OSNRs in the interval from 12 dB to 22 dB. Insert PMD to the polarization-multiplexed signals. Simulate the PMD channel of a 300-km SSMF link with 200 sections and average DGD parameter $\overline{|\tau|} = 0.5$ ps/$\sqrt{\text{km}}$. Consider also transmitter and local oscillator lasers of 50-kHz linewidth, and assume that there is a 500-MHz frequency offset between such lasers. After the optical front-end, send the signals to a low-pass tenth order super-Gaussian filter with 25-GHz bandwidth, and sample the signals at 2 Sa/Symbol. Separate the polarization-multiplexed signals using the CMA, with an 11-tap equalizer. Consider also frequency recovery and Viterbi and Viterbi phase recovery with $N = 40$. Estimate the AIR for the scenarios with SD-SW and SD-BW decoders, assuming in both cases a circularly symmetric Gaussian channel. Discuss the results.

6. Repeat Problem 5 with Nyquist-shaped 16-QAM signals and an OSNR in the interval from 16 dB to 26 dB. Use RDE with the CMA pre-convergence to separate the polarization-multiplexed signals, and perform phase recovery with BPS algorithm. Configure the BPS algorithm with $N = 64$ symbols, and consider $B = 64$ tests phases. Discuss the results.

7. Suppose a non-binary pragmatic FEC scheme that, between OSNRs of 16 dB and 21 dB, needs an extra 0.5 dB gap to guarantee a vanishing error performance, compared to an ideal FEC scheme. Plot the AIR for SD-SW decoders of Problem 6, and add a new curve taking into account the OSNR gap. What is the information rate gap at 18 dB, due to the use of the pragmatic FEC scheme?

8. Suppose a pragmatic binary FEC scheme with the same OSNR gap of 0.5 dB in the OSNR interval from 16 dB to 21 dB. Plot the net data rate for the scenario in Problem 6, supposing ideal SD-BW decoders. Add a new curve, considering the pragmatic binary FEC scheme. Using the pragmatic FEC scheme, what is the

---

[10]In all simulation problems, transmit signals with at least 2^{16} symbols per polarization orientation. Consider optical modulation and coherent detection in the simulation models.

required OSNR to achieve a net data rate of 350 Gb/s? With such OSNR, what is the net data rate achievable for an ideal FEC scheme?

8.5 Matlab/Octave Functions

8.5.1 Functions for Sect. 8.3

Matlab/Octave Code 8.1 Estimation of theoretical AIRs for CM schemes with SD decoding under an AWGN channel

```
function [AIR] = AIR_SDAWGN(SNRdB,ModFormat,AIREval)
%%%%%%%%%%%%%%%%%%%%%%%%%%%%%%%%%%%%%%%%%%%%%%%%%%%%%%%%%%%%%%%%%%%%%%%%%
% AIRAWGN [AIR] = AIR_SDAWGN(SNRdB,ModFormat,AIREval)             %
%                                                                 %
%   This function calculates approximations of the theoretical AIRs for   %
% schemes with SD-SW and SD-BW decoders under the circularly symmetric    %
% Gaussian channel. The approximation is given by a 10-point Gauss-Hermite%
% Quadrature. The supported modulation formats are 'QPSK' and '16QAM'. For%
% SD-BW decoders, the AIR is based on the MI, while for SD-BW decoders,    %
% the AIR is based on the GMI. For SD-BW decoders, Gray labeling is       %
% assumed. The binary labels are equal to the ones used in the function   %
% 'SymbolGeneration'.                                             %
%                                                                 %
% Input:                                                          %
%   SNRdB     = SNR in dB;                                        %
%   ModFormat = Modulation format: 'QPSK' or '16QAM';             %
%   AIREval   = Defines the AIR to the evaluated: 'SD-SW' or 'SD-BW';  %
%                                                                 %
% Output:                                                         %
%   AIR = Approximation of the theoretical AIR (in bit/symbol) for SD-SW %
%         or SD-BW decoders (depending on the variable 'AIREval') under a %
%         circularly symmetric Gaussian channel.                 %
%                                                                 %
% This function is part of the book Digital Coherent Optical Systems;  %
% Darli A. A. Mello and Fabio A. Barbosa;                         %
%%%%%%%%%%%%%%%%%%%%%%%%%%%%%%%%%%%%%%%%%%%%%%%%%%%%%%%%%%%%%%%%%%%%%%%%%
    % Defining parameters according to the modulation format:
    switch ModFormat
        case 'QPSK'
            % Modulation order and number of bits per symbol:
            M = 4 ; m = log2(M);
            % Constellation symbols:
            x = [-1+1i; 1+1i; -1-1i; 1-1i] ; x = x/sqrt(2);
            % Bit-mapping:
            bmap = [0 1; 0 0; 1 1; 1 0];
        case '16QAM'
            % Modulation order and number of bits per symbol:
            M = 16 ; m = log2(M);
            % Constellation symbols:
            x = [3+3i; 1+3i; -1+3i; -3+3i; 3+1i; 1+1i; -1+1i; -3+1i;...
                3-1i; 1-1i; -1-1i; -3-1i; 3-3i; 1-3i; -1-3i; -3-3i];
            x = x/sqrt(10);
            % Bit-mapping:
            bmap = [0 0 0 0; 0 1 0 0; 0 1 0 1; 0 0 0 1; 1 0 0 0;...
                    1 1 0 0; 1 1 0 1; 1 0 0 1; 1 0 1 0; 1 1 1 0;...
                    1 1 1 1; 1 0 1 1; 0 0 1 0; 0 1 1 0; 0 1 1 1; 0 0 1 1];
    end
```

```matlab
% Constants of the 10-point Gauss Hermite Quadrature:
Zeta  = [-3.43615911883773760; -2.53273167423278980; ...
         -1.75668364929988177; -1.03661082978951365; ...
         -0.34290132722370461;  0.34290132722370461; ...
          1.03661082978951365;  1.75668364929988177; ...
          2.53273167423278980;  3.43615911883773760];
Gamma = [0.76404328552326206e-5; 0.13436457467812327e-2; ...
         0.33874394455481063e-1; 0.24013861108231469;     ...
         0.61086263373532580;    0.61086263373532580;     ...
         0.24013861108231469;    0.33874394455481063e-1; ...
         0.13436457467812327e-2; 0.76404328552326206e-5];

% Calculating the standard deviation:
SNR = 10.^(SNRdB/10) ; Stdev = sqrt(mean(abs(x).^2)./(2*SNR));

switch AIREval
case 'SD-SW'
    % MI Calculation (Gauss-Hermite quadrature):
    AIR = zeros(numel(SNRdB),1);
    for n = 1:numel(Stdev)
    Sum = 0;
    for i = 1:M
        for l_1 = 1:numel(Zeta)
            for l_2 = 1:numel(Zeta)
                num = sum(exp(-(abs(x(i)-x).^2 + 2*sqrt(2)*Stdev(n)*...
                    real((Zeta(l_1)+1i*Zeta(l_2)).*(x(i)-x)))/...
                    (2*Stdev(n)^2)));
                Sum = Sum + Gamma(l_1)*Gamma(l_2)*log2(num);
            end
        end
    end
    AIR(n) = m - 1/(M*pi)*Sum;
    end
case 'SD-BW'
    % GMI Calculation (Gauss-Hermite quadrature):
    AIR = zeros(numel(SNRdB),1);
    for n = 1:numel(Stdev)
    Sum = 0;
    for k = 1:m
        for b = 0:1
            xSet = x(bmap(:,m+1-k) == b);
            for i = 1:M/2
                for l_1 = 1:numel(Zeta)
                    for l_2 = 1:numel(Zeta)
                        % Numerator
                        num = sum(exp(-(abs(xSet(i)-x).^2 +...
                            2*sqrt(2)*Stdev(n)*real((Zeta(l_1)+1i*...
                            Zeta(l_2)).*(xSet(i)-x)))/(2*Stdev(n)^2)));
                        % Denominator
                        den = sum(exp(-(abs(xSet(i)-xSet).^2 +...
                            2*sqrt(2)*Stdev(n)*real((Zeta(l_1)+1i*...
                            Zeta(l_2)).*(xSet(i)-xSet)))/...
                            (2*Stdev(n)^2)));
                        Sum = Sum + Gamma(l_1)*Gamma(l_2)*...
                            log2(num/den);
                    end
                end
            end
        end
    end
    AIR(n) = m - 1/(M*pi)*Sum;
    end
end
```

Matlab/Octave Code 8.2 Estimation of AIRs assuming an AWGN channel for non-binary CM schemes with SD-SW decoders

```
function [AIR] = AIR_SDSW(x,y,ModFormat)
%%%%%%%%%%%%%%%%%%%%%%%%%%%%%%%%%%%%%%%%%%%%%%%%%%%%%%%%%%%%%%%%%%%%%%%%%%%%%%
% AIR_SDSW [AIR] = AIR_SDSW(x,y,ModFormat)                                 %
%                                                                         %
%   This function estimates AIRs for CM schemes with soft-decision        %
% symbol-wise (SD-SW) decoders. It is assumed a circularly symmetric      %
% Gaussian channel                                                        %
%                            y = h*x + n                                   %
% where 'y' is the received symbol, 'x' is the transmitted symbol, 'h' is %
% a scalar and 'n' is a complex Gaussian random variable with total       %
% variance TSigma = 2*sigma^2. All input sequences must be synchronized.  %
%                                                                         %
% Input:                                                                  %
%   x         = Sequence of symbols transmitted in one pol. orientation   %
%               (column vector) normalized to unitary power;              %
%   y         = Sequence of symbols received in one pol. orientation      %
%               (column vector) normalized to unitary power;              %
%   ModFormat = Modulation format: 'QPSK' or '16QAM';                     %
%                                                                         %
% Output:                                                                 %
%   AIR = AIR estimate for a CM scheme with SW-SD decoding;               %
%                                                                         %
% This function is part of the book Digital Coherent Optical Systems;     %
% Darli A. A. Mello and Fabio A. Barbosa;                                 %
%%%%%%%%%%%%%%%%%%%%%%%%%%%%%%%%%%%%%%%%%%%%%%%%%%%%%%%%%%%%%%%%%%%%%%%%%%%%%%

    % Defining parameters according to the modulation format:
    switch ModFormat
        case 'QPSK'
            % Modulation order and constellation symbols:
            M = 4 ; s = [-1+1i; 1+1i; -1-1i; 1-1i] ; s = s/sqrt(2);
        case '16QAM'
            % Modulation order and constellation symbols:
            M = 16 ; s = [3+3i; 1+3i; -1+3i; -3+3i; 3+1i; 1+1i; -1+1i; ...
                -3+1i; 3-1i; 1-1i; -1-1i; -3-1i; 3-3i; 1-3i; -1-3i; -3-3i];
            s = s/sqrt(10);
    end

    % Estimating the variance of the complex Gaussian distr.:
    h = x'*y/norm(x)^2 ; Sigma2 = mean(abs(y-h*x).^2)/2;

    % Estimation of the conditional pdf qY|X(y|x):
    qY_X = (1/(2*pi*Sigma2)*exp(-abs(y-h*x).^2/(2*Sigma2)));

    % Estimation of the conditional pdf qY|X(y|s), where 's' represents
    % constellation symbols:
    qY_S = sum((1/(2*pi*Sigma2)*exp(-abs(repmat(y,1,numel(s))-h.*...
        repmat(s.',size(y,1),1)).^2/(2*Sigma2))),2);

    % Estimating the AIR:
    AIR = sum(log2(qY_X./(1/M*qY_S)))/numel(y);
end
```

Matlab/Octave Code 8.3 Estimation of theoretical AIRs for CM schemes with HD decoding under an AWGN channel

```
function [AIR] = AIR_HDAWGN(SNRdB,ModFormat,AIREval)
%%%%%%%%%%%%%%%%%%%%%%%%%%%%%%%%%%%%%%%%%%%%%%%%%%%%%%%%%%%%%%%%%%%%%%%%%%%%
% AIRAWGN [AIR] = AIR_HDAWGN(SNRdB,ModFormat,AIREval)                     %
%                                                                         %
%   This function calculates approximations of the theoretical AIRs for   %
% schemes with HD-SW and HD-BW decoders under the circularly symmetric    %
% Gaussian channel. As symbol error probability (SEP) and bit error       %
% probability (BEP), we use theoretical pre-FEC error performance         %
% expressions. For BEP, we consider an error expression derived for Gray  %
% mapping. The supported modulation formats are 'QPSK' and '16QAM'.       %
%                                                                         %
% Input:                                                                  %
%   SNRdB    = SNR in dB;                                                 %
%   ModFormat = Modulation format: 'QPSK' or '16QAM';                     %
%   AIREval   = Defines the AIR to the evaluated: 'HD-SW' or 'HD-BW';     %
%                                                                         %
% Output:                                                                 %
%   AIR = Approximation of the theoretical AIR (in bit/symbol) for HD-SW  %
%         or HD-BW decoders (depending on the variable 'AIREval') under a %
%         circularly symmetric Gaussian channel.                         %
%                                                                         %
% This function is part of the book Digital Coherent Optical Systems;     %
% Darli A. A. Mello and Fabio A. Barbosa;                                 %
%%%%%%%%%%%%%%%%%%%%%%%%%%%%%%%%%%%%%%%%%%%%%%%%%%%%%%%%%%%%%%%%%%%%%%%%%%%%

    % SNR:
    SNR = 10.^(SNRdB/10);

    % Defining parameters according to the modulation format:
    switch ModFormat
        case 'QPSK'
            % QPSK modulation
            if strcmp(AIREval,'HD-BW')
                BEP = (1/2)*erfc(sqrt(SNR/2));
            elseif strcmp(AIREval,'HD-SW')
                M = 4;
            end
        case '16QAM'
            % 16QAM modulation
            if strcmp(AIREval,'HD-BW')
                BEP = (3/8)*erfc(sqrt(SNR/10)) +...
                    (1/4)*erfc(3*sqrt(SNR/10)) - (1/8)*erfc(5*sqrt(SNR/10));
            elseif strcmp(AIREval,'HD-SW')
                M = 16;
            end
    end
    if strcmp(AIREval,'HD-SW')
        SEP = 2*(sqrt(M)-1)/sqrt(M)*erfc(sqrt(3*SNR/(2*(M-1))))...
            - (((sqrt(M)-1)/sqrt(M))*erfc(sqrt(3*SNR/(2*(M-1))))).^2;
    end

    % AIR estimate:
    switch AIREval
        case 'HD-SW'
            [AIR] = AIR_HDSW(SEP,ModFormat);
        case 'HD-BW'
            [AIR] = AIR_HDBW(BEP,ModFormat);
    end
end
```

Matlab/Octave Code 8.4   Estimation of AIRs assuming an AWGN channel for non-binary CM schemes with HD-SW decoders

```
function [AIR] = AIR_HDSW(SEP,ModFormat)
%%%%%%%%%%%%%%%%%%%%%%%%%%%%%%%%%%%%%%%%%%%%%%%%%%%%%%%%%%%%%%%%%%%%%%%%%%%%%%
% AIR_SWHD [AIR] = AIR_HDSW(SEP,ModFormat)                                  %
%                                                                          %
%    This function estimates AIRs for CM schemes with hard-decision symbol-%
% wise (HD-SW) decoders. It is assumed that the decoding is based on       %
% the Hamming distance, so that we can interpret that the decoder sees a   %
% 'M'-ary symmetric channel with error probability 'SEP', where 'M' is the%
% modulation order. In fact, 'SEP' can be estimated as the pre-FEC symbol  %
% error rate (SER).                                                        %
%                                                                          %
% Input:                                                                   %
%   SEP       = Pre-FEC symbol error probability. 'SEP' can be estimated   %
%                 as the pre-FEC SER (per pol. orientation);               %
%   ModFormat = Modulation format;                                         %
%                                                                          %
% Output:                                                                  %
%   AIR = AIR estimate for a CM scheme with HD-SW decoders;                %
%                                                                          %
% This function is part of the book Digital Coherent Optical Systems;      %
% Darli A. A. Mello and Fabio A. Barbosa;                                  %
%%%%%%%%%%%%%%%%%%%%%%%%%%%%%%%%%%%%%%%%%%%%%%%%%%%%%%%%%%%%%%%%%%%%%%%%%%%%%%

    % Defining parameters according to the modulation format:
    switch ModFormat
        case 'QPSK'
            M = 4;
        case '16QAM'
            M = 16;
    end
    % Binary entropy calculated with 'SEP' and AIR estimate:
    hb  = -SEP.*log2(SEP) - (1-SEP).*log2(1-SEP);
    AIR = log2(M) - hb - SEP*log2(M-1);
end
```

Matlab/Octave Code 8.5   LLR calculation assuming an AWGN channel

```
function L = LLR(y,x,bmap,Sigma2)
%%%%%%%%%%%%%%%%%%%%%%%%%%%%%%%%%%%%%%%%%%%%%%%%%%%%%%%%%%%%%%%%%%%%%%%%%%%%%%
% [L] = LLR(y,x,bmap,TSigma)                                               %
%                                                                          %
%    This function calculated LLRs considering the exact expression for a  %
% circularly symmetric Gaussian channel, for the received symbols 'y',     %
% transmitted symbols 'x', and binary labels given by 'bmap'.              %
%                                                                          %
% Input:                                                                   %
%   y        = Input signal (one pol. orientation) normalized to unitary   %
%                power. 'y' must be a column vector;                       %
%   x        = Reference constellation normalized to unitary power. 'x' must%
%                be a column vector;                                       %
%   bmap     = Binary label of each symbol of the reference constellation  %
%                'x' (in corresponding order to the column vector 'x'). 'bmap'%
%                must be a matrix with 'm = log2(M)' columns and 'M' rows,  %
%                where 'M' is the modulation order;                        %
%   Sigma2   = Estimate of the variance of the circularly symmetric Gaussian%
%                channel. 'Sigma2' is the variance per dimension.          %
%                                                                          %
% Output:                                                                  %
%   L = Estimated LLRs (column vector);                                    %
%                                                                          %
% This function is part of the book Digital Coherent Optical Systems;      %
% Darli A. A. Mello and Fabio A. Barbosa;                                  %
%%%%%%%%%%%%%%%%%%%%%%%%%%%%%%%%%%%%%%%%%%%%%%%%%%%%%%%%%%%%%%%%%%%%%%%%%%%%%%
```

```
    % Number of bits per symbol:
    m = size(bmap,2);

    % LLR estimation:
    L = zeros(numel(y),m);
    for k = 1:m
        % Sets with bit 'b = {0,1}' at position 'k':
        xSet_b1 = x(bmap(:,k)==1) ; xSet_b0 = x(bmap(:,k)==0);

        % LLR estimation:
        num = zeros(numel(y),1) ; den = zeros(numel(y),1);
        for i = 1:m^2/2
            num = num + exp(-abs(y-xSet_b1(i)).^2/(2*Sigma2));
            den = den + exp(-abs(y-xSet_b0(i)).^2/(2*Sigma2));
        end
        L(:,k) = log(num./den);
    end
end
```

Matlab/Octave Code 8.6 Estimation of AIRs assuming an AWGN channel for binary CM schemes with SD-BW decoders

```
function [AIR] = AIR_SDBW(x,b,y,ModFormat)
%%%%%%%%%%%%%%%%%%%%%%%%%%%%%%%%%%%%%%%%%%%%%%%%%%%%%%%%%%%%%%%%%%%%%%%%%
% AIR_SDBW [AIR] = AIR_SDBW(x,b,y,ModFormat)                          %
%                                                                     %
%   This function estimates AIRs for CM schemes with soft-decision bit-wise%
% (SD-BW) decoders. It is assumed a circularly symmetric Gaussian channel:%
%                         y = h*x + n                                 %
% where 'y' is the received symbol, 'x' is the transmitted symbol, 'h' is %
% a scalar and 'n' is a complex Gaussian random variable with total  %
% variance TSigma2 = 2*sigma^2. The AIRs are estimated using LLRs. LLRs %
% are calculated using the exact expression for a circularly symmetric %
% Gaussian channel. Gray mapping is assumed.                          %
%                                                                     %
% Input:                                                              %
%   x        = Sequence of symbols transmitted in one pol. orientation %
%              (column vector) normalized to unitary power;           %
%   b        = Sequence of bits transmitted in one pol. orientation   %
%              (column vector) normalized to unitary power;           %
%   y        = Sequence of symbols received in one pol. orientation   %
%              (column vector);                                       %
%   ModFormat = Modulation format;                                    %
%                                                                     %
% Output:                                                             %
%   AIR = AIR estimate for a CM scheme with BW-SD decoding;           %
%                                                                     %
% This function is part of the book Digital Coherent Optical Systems; %
% Darli A. A. Mello and Fabio A. Barbosa;                             %
%%%%%%%%%%%%%%%%%%%%%%%%%%%%%%%%%%%%%%%%%%%%%%%%%%%%%%%%%%%%%%%%%%%%%%%%%

    % Defining parameters according to the modulation format:
    switch ModFormat
        case 'QPSK'
            % Modulation order and number of bits per symbol:
            M = 4 ; m = log2(M);
            % Bit-mapping:
            bmap = [0 1; 0 0; 1 1; 1 0];
            % Constellation symbols:
            s = [-1+1i; 1+1i; -1-1i; 1-1i].' ; s = s/sqrt(2);
        case '16QAM'
            % Modulation order and number of bits per symbol:
            M = 16 ; m = log2(M);
            % Bit-mapping:
            bmap = [0 0 0 0; 0 1 0 0; 0 1 0 1; 0 0 0 1; 1 0 0 0;...
```

```
                        1 1 0 0; 1 1 0 1; 1 0 0 1; 1 0 1 0; 1 1 1 0;...
                        1 1 1 1; 1 0 1 1; 0 0 1 0; 0 1 1 0; 0 1 1 1; 0 0 1 1];
            % Constellation symbols:
            s = [3+3i; 1+3i; -1+3i; -3+3i; 3+1i; 1+1i; -1+1i; -3+1i;...
                 3-1i; 1-1i; -1-1i; -3-1i; 3-3i; 1-3i; -1-3i; -3-3i].';
            s = s/sqrt(10);
    end

    % Estimating the variance of the complex Gaussian distr.:
    h = x'*y/norm(x)^2 ; Sigma2 = mean(abs(y-h*x).^2)/2;

    % Calculating LLRs:
    L = LLR(y,h*s,bmap,Sigma2);

    % Reshaping the transmitted bits:
    b = reshape(b',m,numel(b)/m)';

    % Computing the AIR (performing minimization over 'a'):
    [~,AIRaux] = fminbnd(@(a) sum(mean(log2(1 + exp(a*(-1).^b.*L))))),0,2);
    AIR        = m - AIRaux;
end
```

Matlab/Octave Code 8.7 Estimation of AIRs assuming an AWGN channel for binary CM schemes with HD-BW decoders

```
function [AIR] = AIR_HDBW(BEP,ModFormat)
%%%%%%%%%%%%%%%%%%%%%%%%%%%%%%%%%%%%%%%%%%%%%%%%%%%%%%%%%%%%%%%%%%%%%%%%%%%%
% AIR_BWHD [AIR] = AIR_HDBW(BEP,ModFormat)                                %
%                                                                        %
%    This function estimates AIRs for CM schemes with hard-decision bit- %
% wise (HD-BW) decoders. It is assumed that the decoding is based on     %
% the Hamming distance, so that we can interpret that the decoder sees 'm'%
% parallel binary symmetric channels with error probability 'BEP', where %
% 'm = log2(M)' and 'M' is the modulation order. In fact, 'BEP' is the   %
% pre-FEC average error probability across the 'm' bit positions. 'BEP'  %
% can be estimated as the pre-FEC bit error rate (BER).                  %
%                                                                        %
% Input:                                                                 %
%    BEP      = Pre-FEC average bit error probability across the 'm' bit %
%               positions. 'BEP' can be estimated as the pre-FEC BER (per %
%               pol. orientation);                                       %
%    ModFormat = Modulation format: 'QPSK' or '16QAM';                   %
%                                                                        %
% Output:                                                                %
%    AIR = AIR estimate for a CM scheme with HD-BW decoding;             %
%                                                                        %
% This function is part of the book Digital Coherent Optical Systems;    %
% Darli A. A. Mello and Fabio A. Barbosa;                                %
%%%%%%%%%%%%%%%%%%%%%%%%%%%%%%%%%%%%%%%%%%%%%%%%%%%%%%%%%%%%%%%%%%%%%%%%%%%%

    % Defining parameters according to the modulation format:
    switch ModFormat
        case 'QPSK'
            M = 4;
        case '16QAM'
            M = 16;
    end
    % Binary entropy calculated with 'BEP' and AIR estimate:
    hb  = -BEP.*log2(BEP) - (1-BEP).*log2(1-BEP);
    AIR = log2(M)*(1 - hb);
end
```

References

1. D. Mello, E. Offer, J. Reichert, Error arrival statistics for FEC design in four-wave mixing limited systems, in *Proceedings of Optical Fiber Communication Conference (OFC)* (Optical Society of America, America, 2003), p. ThN3
2. G. Liga, A. Alvarado, E. Agrell, P. Bayvel, Information rates of next-generation long-haul optical fiber systems using coded modulation. J. Lightwave Technol. **35**(1), 113–123 (2017)
3. L. Schmalen, A. Alvarado, R. Rios-Müller, Performance prediction of nonbinary forward error correction in optical transmission experiments. J. Lightwave Technol. **35**(4), 1015–1027 (2017)
4. A. Sheikh, A.G.i. Amat, G. Liva, Achievable information rates for coded modulation with hard decision decoding for coherent fiber-optic systems. J. Lightwave Technol. **35**(23), 5069–5078 (2017)
5. G. Ungerboeck, Channel coding with multilevel/phase signals. IEEE Trans. Inf. Theory **28**(1), 55–67 (1982)
6. H. Imai, S. Hirakawa, A new multilevel coding method using error-correcting codes. IEEE Trans. Inf. Theory **23**(3), 371–377 (1977)
7. I.B. Djordjevic, B. Vasic, Nonbinary LDPC codes for optical communication systems. IEEE Photon. Technol. Lett. **17**(10), 2224–2226 (2005)
8. E. Zehavi, 8-PSK trellis codes for a Rayleigh channel. IEEE Trans. Commun. **40**(5), 873–884 (1992)
9. A. Alvarado, E. Agrell, D. Lavery, R. Maher, P. Bayvel, Replacing the soft-decision FEC limit paradigm in the design of optical communication systems. J. Lightwave Technol. **33**(20), 4338–4352 (2015). http://jlt.osa.org/abstract.cfm?URI=jlt-33-20-4338
10. A. Leven, F. Vacondio, L. Schmalen, S. ten Brink, W. Idler, Estimation of soft FEC performance in optical transmission experiments. IEEE Photon. Technol. Lett. **23**(20), 1547–1549 (2011)
11. I.B. Djordjevic, H.G. Batshon, L. Xu, T. Wang, Coded polarization-multiplexed iterative polar modulation (PM-IPM) for beyond 400 Gbps serial optical transmission, in *Proceedings of Optical Fiber Communication Conference (OFC)* (Optical Society of America, America, 2010), p. OMK2. http://www.osapublishing.org/abstract.cfm?URI=OFC-2010-OMK2
12. X. Liu, S. Chandrasekhar, T. Lotz, P. Winzer, H. Haunstein, S. Randel, S. Corteselli, B. Zhu, D. Peckham, Generation and FEC-decoding of a 231.5-Gb/s PDM-OFDM signal with 256-iterative-polar-modulation achieving 11.15-b/s/Hz intrachannel spectral efficiency and 800-km reach, in *Proceedings of National Fiber Optic Engineers Conference (NFOEC)* (Optical Society of America, America, 2012), p. PDP5B.3. http://www.osapublishing.org/abstract.cfm?URI=NFOEC-2012-PDP5B.3
13. M.P. Yankov, D. Zibar, K.J. Larsen, L.P.B. Christensen, S. Forchhammer, Constellation shaping for fiber-optic channels with QAM and high spectral efficiency. IEEE Photon. Technol. Lett. **26**(23), 2407–2410 (2014)
14. T. Fehenberger, G. Böcherer, A. Alvarado, N. Hanik, LDPC coded modulation with probabilistic shaping for optical fiber systems, in *Proceedings of Optical Fiber Communication Conference (OFC)* (Optical Society of America, America, 2015), p. Th2A.23. http://www.osapublishing.org/abstract.cfm?URI=OFC-2015-Th2A.23
15. C.E. Shannon, A mathematical theory of communication. SIGMOBILE Mobile Comput. Commun. Rev. **5**(1), 3–55 (2001). http://doi.acm.org/10.1145/584091.584093
16. J. Cho, P.J. Winzer, Probabilistic constellation shaping for optical fiber communications. J. Lightwave Technol. **37**(6), 1590–1607 (2019)
17. D.J. Costello, G.D. Forney, Channel coding: the road to channel capacity. Proc. IEEE **95**(6), 1150–1177 (2007)
18. M. Arabaci, I.B. Djordjevic, L. Xu, T. Wang, Nonbinary LDPC-coded modulation for rate-adaptive optical fiber communication without bandwidth expansion. IEEE Photon. Technol. Lett. **24**(16), 1402–1404 (2012)

19. A. Sheikh, A.G. Amat, M. Karlsson, Nonbinary staircase codes for spectrally and energy efficient fiber-optic systems, in *Proceedings of Optical Fiber Communication Conference (OFC)* (2017), pp. 1–3
20. A. Alvarado, E. Agrell, Achievable rates for four-dimensional coded modulation with a bit-wise receiver, in *Proceedings of Optical Fiber Communication Conference (OFC)*. (Optical Society of America, America, 2014), p. M2C.1. http://www.osapublishing.org/abstract.cfm?URI=OFC-2014-M2C.1
21. A. Martinez, A. Guillen i Fabregas, G. Caire, F.M.J. Willems, Bit-interleaved coded modulation revisited: a mismatched decoding perspective. IEEE Trans. Inf. Theory **55**(6), 2756–2765 (2009)
22. U. Wachsmann, R.F.H. Fischer, J.B. Huber, Multilevel codes: theoretical concepts and practical design rules. IEEE Trans. Inf. Theory **45**(5), 1361–1391 (1999)
23. T.M. Cover, J.A. Thomas, *Elements of Information Theory (Wiley Series in Telecommunications and Signal Processing)* (Wiley, Hoboken, 2006)
24. T.K. Moon, *Error Correction Coding: Mathematical Methods and Algorithms* (Wiley, USA, 2005)
25. D.M. Arnold, H. . Loeliger, P.O. Vontobel, A. Kavcic, W. Zeng, Simulation-based computation of information rates for channels with memory. IEEE Trans. Inf. Theory **52**(8), 3498–3508 (2006)
26. E. Agrell, M. Secondini, Information-theoretic tools for optical communications engineers, in *Proceedings of IEEE Photonics Conference (IPC)* (2018), pp. 1–5
27. A. Alvarado, T. Fehenberger, B. Chen, F.M.J. Willems, Achievable information rates for fiber optics: applications and computations. J. Lightwave Technol. **36**(2), 424–439 (2018)
28. F.W. Olver, D.W. Lozier, R.F. Boisvert, C.W. Clark, *NIST Handbook of Mathematical Functions*, 1st edn. (Cambridge University, Cambridge, 2010)
29. T. Fehenberger, A. Alvarado, P. Bayvel, N. Hanik, On achievable rates for long-haul fiber-optic communications. Opt. Express **23**(7), 9183–9191 (2015). http://www.opticsexpress.org/abstract.cfm?URI=oe-23-7-9183
30. F. Buchali, F. Steiner, G. Böcherer, L. Schmalen, P. Schulte, W. Idler, Rate adaptation and reach increase by probabilistically shaped 64-QAM: an experimental demonstration. J. Lightwave Technol. **34**(7), 1599–1609 (2016). http://jlt.osa.org/abstract.cfm?URI=jlt-34-7-1599
31. N. Merhav, G. Kaplan, A. Lapidoth, S. Shamai Shitz, On information rates for mismatched decoders. IEEE Trans. Inf. Theory **40**(6), 1953–1967 (1994)
32. A. Sheikh, On hard-decision forward error correction with application to high-throughput fiber-optic communications. Ph.D. dissertation (Chalmers University of Technology, Gothenburg, 2019). https://research.chalmers.se/en/publication/510548
33. G. Böcherer, Principles of coded modulation. Habilitation thesis (Technical University of Munich, Munich, 2018). http://www.georg-boecherer.de/bocherer2018principles.pdf
34. F. Steiner, Coding for higher-order modulation and probabilistic shaping. Ph.D. dissertation (Technical University of Munich, Munich, 2020). http://mediatum.ub.tum.de/?id=1520127
35. A. Goldsmith, *Wireless Communications* (Cambridge University, New York, 2005)
36. L. Szczecinski, A. Alvarado, *Bit-interleaved coded modulation: fundamentals, analysis, and design* (Wiley, Chichester, 2015)
37. A. Alvarado, D.J. Ives, S.J. Savory, P. Bayvel, On the impact of optimal modulation and FEC overhead on future optical networks. J. Lightwave Technol. **34**(9), 2339–2352 (2016). http://jlt.osa.org/abstract.cfm?URI=jlt-34-9-2339
38. K. Cho, D. Yoon, On the general BER expression of one- and two-dimensional amplitude modulations. IEEE Trans. Commun. **50**(7), 1074–1080 (2002)

Index

© Springer Nature Switzerland AG 2021
D. A. de Arruda Mello, F. A. Barbosa, *Digital Coherent Optical Systems*,
Optical Networks, https://doi.org/10.1007/978-3-030-66541-8

Printed in the United States
by Baker & Taylor Publisher Services